Particulate Drying

In the process industry, understanding the unit operation of particulate drying is imperative to yield products with desired properties and characteristics and to ensure process safety, optimal energy efficiency and drying performance, as well as low environmental impact. There are many techniques and tools available, which can cause confusion. *Particulate Drying: Techniques and Industry Applications* provides an overview of various particulate drying techniques, their advantages and limitations, industrial applications, and simple design methods.

This book:

- Covers advances in particulate drying and their importance in the process industry
- Highlights recent developments in conventional drying techniques and new drying technologies
- Helps readers gain insight into selecting the appropriate drying techniques for a particular product
- Summarizes various applications from a wide range of industries, including chemical, food, pharmaceutical, biotech, polymer, mineral, and agro-industries
- Projects future research trends and demands in particulate drying

This book serves as a reference for process and plant engineers as well as researchers in the fields of particulate processing, mineral processing, food processing, chemical engineering, and mechanical engineering, especially those involved in the selection of drying equipment for particulate solids and R&D of drying of particulate materials.

Advances in Drying Science and Technology

Series Editor: Arun S. Mujumdar, McGill University, Quebec, Canada

Industrial Heat Pump-Assisted Wood Drying
Vasile Minea

Intelligent Control in Drying
Alex Martynenko and Andreas Bück

Drying of Biomass, Biosolids, and Coal: For Efficient Energy Supply and Environmental Benefits
Shusheng Pang, Sankar Bhattacharya, Junjie Yan

Drying and Roasting of Cocoa and Coffee
Ching Lik Hii and Flavio Meira Borem

Heat and Mass Transfer in Drying of Porous Media
Peng Xu, Agus P. Sasmito, and Arun S. Mujumdar

Freeze Drying of Pharmaceutical Products
Davide Fissore, Roberto Pisano, and Antonello Barresi

Frontiers in Spray Drying
Nan Fu, Jie Xiao, Meng Wai Woo, Xiao Dong Chen

Drying in the Dairy Industry
Cécile Le Floch-Fouere, Pierre Schuck, Gaëlle Tanguy, Luca Lanotte, Romain Jeantet

Spray Drying Encapsulation of Bioactive Materials
Seid Mahdi Jafari and Ali Rashidinejad

Flame Spray Drying: Equipment, Mechanism, and Perspectives
Mariia Sobulska and Ireneusz Zbicinski

Advanced Micro-Level Experimental Techniques for Food Drying and Processing Applications
Azharul Karim, Sabrina Fawzia and Mohammad Mahbubur Rahman

Mass Transfer Driven Evaporation of Capillary Porous Media
Rui Wu, Marc Prat

Particulate Drying: Techniques and Industry Applications
Sachin Vinayak Jangam, Chung Lim Law, and Shivanand Shankarrao Shirkole

For more information about this series, please visit: www.routledge.com/Advances-in-Drying-Science-and-Technology/book-series/CRCADVSCITEC

Particulate Drying
Techniques and Industry Applications

Edited by
Sachin Vinayak Jangam
Chung Lim Law
Shivanand Shankarrao Shirkole

CRC Press
Taylor & Francis Group
Boca Raton London New York

CRC Press is an imprint of the
Taylor & Francis Group, an **informa** business

First edition published 2024
by CRC Press
6000 Broken Sound Parkway NW, Suite 300, Boca Raton, FL 33487-2742

and by CRC Press
4 Park Square, Milton Park, Abingdon, Oxon, OX14 4RN

CRC Press is an imprint of Taylor & Francis Group, LLC

ISBN: 978-1-032-07467-2 (hbk)
ISBN: 978-1-032-07473-3 (pbk)
ISBN: 978-1-003-20710-8 (ebk)

DOI: 10.1201/9781003207108

Typeset in Times
by MPS Limited, Dehradun

Contents

Advances in Drying Science and Technology

It is well known that the unit operation of drying is a highly energy-intensive operation encountered in diverse industrial sectors ranging from agricultural processing, ceramics, chemicals, minerals processing, pulp and paper, pharmaceuticals, coal polymer, food, forest products industries, as well as waste management. Drying also determines the quality of the final dried products. The need to make drying technologies sustainable and cost effective via application of modern scientific techniques is the goal of academic as well as industrial R&D activities around the world.

Drying is a truly multi- and interdisciplinary area. Over the last four decades, the scientific and technical literature on drying has seen exponential growth. The continuously rising interest in this field is also evident from the success of numerous international conferences devoted to drying science and technology.

The establishment of this new series of books entitled *Advances in Drying Science and Technology* is designed to provide authoritative and critical reviews and monographs focusing on current developments as well as future needs. It is expected that books in this series will be valuable to academic researchers as well as industry personnel involved in any aspect of drying and dewatering.

The series will also encompass themes and topics closely associated with drying operations, e.g., mechanical dewatering, energy savings in drying, environmental aspects, life-cycle analysis, technoeconomics of drying, electrotechnologies, control and safety aspects, and so on.

ABOUT THE SERIES EDITOR

Dr. Arun S. Mujumdar is an internationally acclaimed expert on drying science and technologies. He was the founding chair in 1978 of the International Drying Symposium (IDS) series and editor in chief of *Drying Technology: An International Journal* since 1988. The fourth enhanced edition of his *Handbook of Industrial Drying,* published by CRC Press, has just appeared. He is the recipient of numerous international awards, including honorary doctorates from Lodz Technical University, Poland, and University of Lyon, France.

Please visit www.arunmujumdar.com for further details.

Series Editor
Dr. Arun S. Mujumdar

Preface

In process industries, the finished products or intermediate products are often in the form of particulate solids and there is a need to remove water or solvent from such particulate solids. Hence, the drying of particulate materials is an important unit operation to the process industry. As drying is an energy-intensive unit operation, an improper selection of drying methods may lead to undesirable consequences. It is important for the process industry, process engineer, and design engineer to become familiar with the various types of drying technologies that are suitable for drying particulate materials. It is equally important to know the advantages and limitations of diverse dryer types and the operating conditions so that involved experts can use good judgement in dryer selection.

This book is dedicated to the fundamentals of particulate drying methods and their industrial applications. It aims to provide readers with an overview of various types of drying techniques that are suitable for particulate drying. Drying techniques that are covered in this book include vibrated bed drying and impinging streams drying, as well as common particulate dryers techniques such as rotary, superheated steam, and flash dryers. Some of the special features of this book include:

- Explaining advances in particulate drying techniques and their importance in the process industry
- Critical reviews of recent developments in conventional and new drying technologies for particulate drying
- Access to selected industrial applications of particulate drying

The first chapter provides a short introduction to drying particulates. This is followed by a chapter on one of the commonly used particulate drying techniques: vibrated bed dryer. Fixed bed drying is the next topic included in this book. This chapter covers fundamentals of fixed bed dryers, variants, and recent attempts to fixed bed drying strategy. It is followed by a chapter on another important type of particulate dryers: pneumatic conveying drying. The fifth chapter is on particulate drying with impinging streams.

The second half of the book starts with a chapter on drying suspensions and pastes in particulate beds. This chapter starts with an introduction to the drying mechanisms using inert beds, various configurations used, and a short discussion of relevant industrial and academia perspectives. A chapter on another traditional industrial drying method, rotary dryers, follows. This chapter not only covers fundamentals, but also numerical simulations with a discussion on industrial applications. The next chapter is on prospects for the development of industrial process for drying nanoformulations. The ninth chapter is on superheated steam drying (SSD) of particulates. This chapter starts with an introduction to SSD, advantages/limitations, an overview of experimental and theoretical studies on SSD, and finally an industrial perspective. The final chapter includes miscellaneous techniques such as spouted bed drying, air impingement drying, screw conveyor drying, mechanically agitated dryers, heat pump drying, and microwave drying.

Overall, this book is designed as a reference source for process engineers, process design engineers, and plant engineers, as well as researchers in the fields of particulate processing, mineral processing, and food processing, as well as chemical engineering and mechanical engineering, especially those who are involved in the selection of drying equipment for particulate solids drying as well as the research and development of drying of particulate materials. This book gathers the expertise of internationally recognized authors who have provided the latest developments in technologies for drying particulates, including updates from recent research works. Primarily, it is suitable for undergraduate and postgraduate students and researchers and industrial practitioners, as well as consultants from various domains.

We are grateful to the authors as well as reviewers who have contributed their valuable time as well as their experience and expertise in the production of this book.

The Editors

Sachin Vinayak Jangam

Chung Lim Law

Shivanand Shankarrao Shirkole

Author/Editor Biographies

Sachin Vinayak Jangam is a senior lecturer in the Department of Chemical and Biomolecular Engineering at the National University of Singapore (NUS). He completed his PhD in chemical engineering at the Institute of Chemical Technology, Mumbai, India. He worked on mathematical modeling and experimental analysis of industrial drying of various products as a major part of his PhD thesis. He then worked as a research fellow at the Minerals, Metals and Materials Technology Center at NUS, developing cost-effective drying techniques for minerals. Sachin has published several research articles, review papers, and book chapters on drying and related fields. He is coauthor of a book on foundational concepts of chemical engineering and has edited several free e-books on drying. Sachin's current work focuses on drying, energy minimization, and pedagogy in chemical engineering education. He has been an assistant editor of the archival journal *Drying Technology* (Taylor & Francis) since 2015.

Chung Lim Law is a professor and head of the Department of Chemical and Environmental Engineering, University of Nottingham Malaysia. He received his PhD in 2004 from the National University of Malaysia and a Post Graduate Certificate of Higher Education (PGCHE) in 2010. Chung Lim has more than 20 years of experience in research and development in drying, especially particulate drying and low-temperature drying. Over the years, he has conducted research works on topics related to preservation and retention of bioactive ingredients of dehydrated bioproducts, food products, and herbal products. His research also focuses on hybrid drying, two-stage drying fluidized bed drying, heat pump drying, and intermittent drying. From 2014 to 2018, he was an assistant and associate editor of the archival journal *Drying Technology* (Taylor & Francis).

Shivanand Shankarrao Shirkole is an assistant professor in the Department of Food Engineering and Technology, Institute of Chemical Technology Mumbai, ICT-IOC Odisha Campus, Bhubaneswar, India. He received his PhD in 2020 from the Department of Food Process Engineering, National Institute of Technology, Rourkela, India. He has more than four years of industrial experience as a plant engineer at a Hyderabad-based Multi-Crop Seed Conditioning Plant. His broad areas of research are low-moisture food safety and thermal processing of food. Presently, he is working on sustainable technologies for food and agri-products. He is an executive committee member of the Association of Food Scientists & Technologists (India), NIT Rourkela chapter, and works as an associate editor for the journal *Drying Technology* (Taylor & Francis).

Contributors

Mukund Haribhau Bade
Department of Mechanical Engineering
Sardar Vallabhbhai National Institute of Technology
Surat, Gujarat, India

Marcos Antonio de Souza Barrozo
School of Chemical Engineering
Federal University of Uberlândia
Uberlândia, Brazil

Ronaldo Correia de Brito
Department of Chemical Engineering
Federal University of São Carlos
São Carlos, SP, Brazil

Sakamon Devahastin
Department of Food Engineering
Faculty of Engineering
King Mongkut's University of Technology Thonburi
Bangkok, Thailand
and
The Academy of Science
The Royal Society of Thailand
Bangkok, Thailand

Masoud Dorfeshan
Department of Mechanical Engineering
Behbahan Khatam Alanbia University of Technology
Behbahan, Iran

Cláudio Roberto Duarte
School of Chemical Engineering
Federal University of Uberlândia
Uberlândia, Brazil

Maria do Carmo Ferreira
Department of Chemical Engineering
Federal University of São Carlos
São Carlos, SP, Brazil

Fábio Bentes Freire
Department of Chemical Engineering
Federal University of São Carlos
São Carlos, SP, Brazil
and
Department of Civil Engineering
Federal University of Technology Curitiba
PR, Brazil

Flavio Bentes Freire
Department of Civil Engineering
Federal University of Technology – Paraná
Curitiba, Brazil

José Teixeira Freire
Department of Chemical Engineering
Federal University of São Carlos
São Carlos, SP, Brazil

Sachin Vinayak Jangam
Department of Chemical and Biomolecular Engineering
National University of Singapore
Singapore

Eknath Kole
Department of Pharmaceutical Technology
University Institute of Chemical Technology
Kavayitri Bahinabai Chaudhari North Maharashtra University Jalgaon
India

Chung Lim Law
Department of Chemical and Environmental Engineering
University of Nottingham Malaysia
Semenyih, Selangor, Malaysia

Salem Mehrzad
Department of Mechanical Engineering, Engineering Faculty
Shahid Chamran University of Ahvaz
Ahvaz, Iran

Lucas Meili
Laboratory of Processes, Center of Technology
Federal University of Alagoas
Maceió, AL, Brazil

Arun Sadashiv Mujumdar
Department of Bioresource Engineering Macdonald College
McGill University
Ste-Anne-de-Bellevue, Quebec, Canada

Jitendra Naik
Department of Pharmaceutical Technology
University Institute of Chemical Technology
Kavayitri Bahinabai Chaudhari North Maharashtra University Jalgaon
India

Sullen Mendonça Nascimento
Department of Engineering
Federal University of Lavras
Lavras, Brazil

Sagar Pardeshi
Department of Pharmaceutical Technology
University Institute of Chemical Technology
Kavayitri Bahinabai Chaudhari North Maharashtra University
Jalgaon, India

Sanjay Kumar Patel
Department of Mechanical Engineering
JSPM's Rajarshi Shahu College of Engineering
Pune, Maharashtra, India

Hugo Perazzini
Institute of Natural Resources
Federal University of Itajubá
Itajubá, MG, Brazil

Maisa Tonon Bitti Perazzini
Institute of Natural Resources
Federal University of Itajubá
Itajubá, MG, Brazil

Somkiat Prachayawarakorn
Department of Chemical Engineering
Faculty of Engineering
King Mongkut's University of Technology Thonburi
Bangkok, Thailand

Dyrney Araujo dos Santos
Institute of Chemistry
Federal University of Goiás
Goiânia, Brazil

Shivanand Shankarrao Shirkole
Department of Food Engineering and Technology
Institute of Chemical Technology Mumbai, ICT – IOC Odisha Campus
Bhubaneswar, India

Somchart Soponronnarit
Division of Energy Technology
School of Energy, Environment and Materials
King Mongkut's University of Technology Thonburi
Bangkok, Thailand
and
The Academy of Science
The Royal Society of Thailand
Bangkok, Thailand

Yang Tao
College of Food Science and Technology
Nanjing Agricultural University
Nanjing, Jiangsu, China

Hong-Wei Xiao
College of Engineering
China Agricultural University
Beijing, China

1 Introduction to Particulate Drying

Chung Lim Law
Department of Chemical and Environmental Engineering, University of Nottingham Malaysia, Semenyih, Selangor, Malaysia

Shivanand Shankarrao Shirkole
Department of Food Engineering and Technology, Institute of Chemical Technology Mumbai, ICT – IOC Odisha Campus, Bhubaneswar, India

Sachin Vinayak Jangam
Department of Chemical and Biomolecular Engineering, National University of Singapore, Singapore

CONTENTS

1.1 PARTICULATE DRYING: FUNDAMENTALS

The drying process converts a wet feedstock into a solid product by evaporation of the liquid (mostly water, but not limited to water) into a vapor phase via application of heat. The feedstock to dryer and be in various physical forms such as solid, semi-solid, or liquid. It is well known that the dryer throughput may vary considerably (from 1 kg/h to hundreds of tonnes of material per hour). Drying is a highly energy-intensive unit operation found in most industrial sectors and accounts for between 10%–20% of national industrial energy consumption in developed countries (Mujumdar, 2009; Mujumdar, 2011; Lee, Jangam and Mujumdar, 2013). Besides this, it involves transient transport phenomena subject to rather severe constraints imposed by material science. The above-mentioned facts help us appreciate the importance of understanding the drying process, and the need for research and development in drying (Mujumdar, 2009; Mujumdar, 2014; Tsotsas and Mujumdar, 2014).

DOI: 10.1201/9781003207108-1

A large fraction of products processed industrially are in particulate form so that drying of wet particulates is of great industrial interest. It is important that we use efficient drying methods to produce engineered dry particulates of desired quality at minimum cost, and with minimal impact on the environment. In order to reduce investment costs, one needs to enhance drying rates within limits imposed by the product properties and end product quality requirements. Several novel gas-particle contactors have been evaluated for drying. Combined modes of heating and hybrid dryers can improve drying performance in some cases. Recent interest in production of nanoparticles by wet processing also has stimulated interest in drying to produce nanoparticles (Lee et al., 2013). Drying of heat-sensitive biotech and pharmaceutical products also pose new challenges. This chapter attempts to provide a general introduction to drying of particulate materials, a brief discussion on classification and selection of dryers, and the importance of particulate drying in industries. A capsule overview is presented of recent developments, including enhancements in conventional drying technologies as well as more innovative new technologies for particulate drying. When the feedstock is in a liquid or sludge form, some sort of pre-treatment is often used to reduce the load on the drying unit. However, this chapter of the book does not cover this aspect. Readers may refer to the latest edition of the handbook of industrial drying for further details (Mujumdar, 2009; Mujumdar, 2011; Mujumdar, 2014).

1.2 INDUSTRIAL IMPORTANCE OF PARTICULATE DRYING

Drying is very common in our daily life as well as in various process industries. Drying may be applied to the processing of raw material or intermediate product into a finished product. Very often, the raw material or the intermediate product or finished product are in the form of particulate solids; hence, drying of particulate solids becomes a very common processing step that we need to deal with. Indeed, it is one of the major unit operations in the particulate solids processing industry that contributes to a whopping 67% of the global industrial output. Table 1.1 shows the various examples of particulate solids that require drying in some processing stage, ranging from grain, food and fruit, to chemicals, pharmaceutical chemicals, active ingredients, extract powders, minerals, ceramics and composites, industrial sludge, saw dust, and wood chips.

All the products listed in Table 1.1 are in the form of particulate solids; they may be fine powders, granules, pellets, or a larger size object of arbitrary shape. The particulate form facilitates free flow of materials that can be transported, packed, stored, dosed, or agglomerated. Therefore, there is a need to perform drying to reduce the moisture content in order to facilitate the free flow of the particulate solids.

For certain products, there are regulations and statutory limits stipulated in legislation or statutory instruments of respective countries. For instance, U.K. statutory instruments 1998 No. 141 Food, the Bread and Flour Regulations 1998, Schedule 1 Essential Ingredients of Flour states that the volatile matter shall not be more than 1%. Further, legislation in some countries puts specific limitations on the maximum allowable moisture content that is legally allowed in certain food products. For instance, cheddar cheese shall have no more than 40% moisture content.

TABLE 1.1
Industry Sectors that Deal with Particulate Drying

Industry sector	Examples
Grain	Rough rice, corn, barley, wheat, oat, sorghum, rye
Food and fruit	Fruit chips, dairy products (milk, whey, creamers), coffee, coffee surrogates, tea, flavours, powdered drinks processed cereal-based foods, potatoes, starch derivatives, sugar beet pulp, fruits, vegetables, spices
Chemicals, fine chemicals, specialty chemicals	Organic chemicals, inorganic chemicals plastic chemicals, resins, plastics, polymers, synthetics, catalysts (zeolite), dyes and pigments, soap and detergent, explosives chemicals, bromine chemicals, catalyst chemicals, consumer chemicals (detergents), nuclear chemicals
Pharmaceutical chemicals	Paracetamol (Acetaminophen), dicalcium phosphate, glucosamine HCl, valine
Bio-pharmaceuticals	Enzymes, pro and/or prebiotics, oils, vitamins, minerals, phytonutrients or nutraceuticals
Agro-chemicals	pesticides, herbicides
Plant based ingredients	Natural flavors, natural colorants, food preservers and enhancers, texturants, antioxidants, natural extract powders, natural bioactive ingredients
Mineral	Granite, dolomite, basalt, quartz sand, feldspar, coke, coal, salts, potash, silica sand, slag sand, Frac sand, kaolin, aggregate, bauxite, boiler ash, calcium carbonate, chalk
Cement	Cement raw materials such as limestone, shale, gypsum
Ceramics	Glass ceramics, silicon carbide, titanium carbide, barium titanate
Composites	Spherulites, semicrystalline polymers, bricks
Industrial sludge	Industrial wastewater sludge, sewage sludge
Wood and lumber	Sawdust, wood chips

Particulate solids drying is important from the point of view of product quality as moisture content of the particulate solids should be kept within an acceptable range. This is especially important to food products, bio-origin products, agricultural products, and biotechnical products. If the moisture content is too high, it is susceptible to microbial growth, which affects the shelf life and may result in degradation of bioactive ingredients that negatively impact the product's nutritional values. If the moisture content is too low, it means that drying is over-done and hence wastes energy and operating costs. This may also accompany the loss of valuable nutrients.

Drying of particulate solids may be required due to requirements of subsequent process steps; for instance, milling of wheat, where flour in the form of fine powder is produced from the milling process. Manufacturing of pharmaceutical tablets requires the conversion of a drug into powder form is required by the tableting process to make it into tablet form. Extraction of bioactive ingredients from bio-origin products may require the fresh bio-origin products to be dried and ground into powder before it is subjected to an extraction process. Rubber chemicals are required to reduce the moisture content before they are used in the vulcanization process of tires.

In certain processing sectors, drying is performed for the purpose of achieving a desirable shape, typically a spherical shape of the solid product. For instance, spray drying is applied to produce hard and dry FCC spheres of a 80 μm diameter. In certain sectors, drying is a useful engineering tool to shape and size the processed products to the desired shape, size, texture, porosity, colour, appearance, surface characteristics, etc.

1.3 TYPICAL ISSUES WITH PARTICULATE DRYING

Drying often needs to fulfil a number of constraints ranging from product quality (which is the most important priority), economics (associated to processing cost), process safety, food safety, fire and explosion hazard, toxicity hazard, handling of solids, transport and flowability of particulate solids, storage and rehydration of dry solids, and finally drying performance (Mujumdar, 2014).

The following are some of the aspects that are relevant to particulate product quality:

- Final moisture content and equilibrium moisture content
- Particle size and size distribution, dispersibility, and dust content
- Flow characteristics
- Textural attributes such as hardness
- Appearance attributes such as colour, size, and shape
- Odour, taste
- Dissolution/rewetting behaviour
- Availability or presence of functional activity of bioactive ingredients
- Caking tendency
- Segregation of originally dissolved components

Particulate solids may be allowed to contain some residual moisture content according to the sale and purchase contract; therefore, there is no need to over-dry particulate solids. Further, even if the particulate solids are completely dry, they may absorb moisture during storage and attain the equilibrium moisture content. It is important to identify the desirable moisture content that is preferably close to the equilibrium moisture content at a storage condition. Hence, over-drying can be avoided and therefore minimise waste in energy utilisation for drying. For the processing industry, it is a challenge in determining the desirable final moisture content because most industrial players are not aware of the importance of finding the suitable final moisture content as well as equilibrium moisture content in a storage condition. In addition, residual moisture content may influence the transport and storage behaviour of the particulate solids as well as their functionality. This is explained in the subsequent paragraphs.

Particle-form solids in bulk may appear in a wide range of particle size distributions. Handling wide particle size distributions of bulk solids may give rise to the problem of segregation, where fine powder and coarse particles tend to segregate during the handling and transportation; hence, causing non-uniformity in terms of particle size and composition. If the non-uniform particulate solids are subjected to drying, it may give rise to the issue of uneven drying performance and, hence, non-uniform product quality. Moreover, when such particulate solids are subjected to fluidised bed drying, fine powders tend to be carried over by the fluidising gas. When they are subjected to horizontal pneumatic conveying after drying, fine powder or small particles tend to settle at the bottom of the pipeline and flow intermittently.

Particulate solids, especially fine powders, have a strong surface cohesion force, causing this type of fine powder to be sticky. Due to the stickiness, the fine powders agglomerate and lump together, causing them to be difficult to handle and dry when they are subjected to drying, especially drying methods that depend on convective drying. The contact efficiency between the drying air and the particulate agglomerates is poor and it reduces the drying efficiency. Agglomeration of fine powders also results in poor heat and mass transfer; therefore, resulting in poor performance of drying and thus high drying operating costs.

Bio-origin products such as grain, fruits, and agricultural products possess skin or exocarp, or hull, which can impede the penetration of water vapour or diffusion of moisture; thus, making drying difficult and takes a lot longer to accomplish. In addition, for some fruits and food products, rapid drying may cause case hardening of the product surface, forming a crusty layer on the surface of the product, hence giving the same effect as exocarp to a bio-origin product mentioned above. Case hardening also results in a puffing effect, which has been reported by many researchers.

During drying, water migrates from the internal to the particulate solid's surface. Depending on the type of drying technique applied, the removal of internal moisture may result in the formation of pores within the solid matrix or collapse of solid matrix due to rapid removal of moisture. For the latter case, it causes noticeable shrinkage due to mechanical stress during drying, especially if the drying condition is harsh, i.e., high temperature or rapid drying, etc. This results in a poorly dehydrated product quality because it has low porosity and poor rehydration ability.

Some particulate solids are hygroscopic and can easily absorb moisture and rehydrate during storage. Hence, this poses additional challenges to the handling and storing of this type of hygroscopic particulate solid.

Many particulate solids, especially those that are biological in origin and those that contain heat-sensitive bioactive ingredients or thermally labile components, are susceptible to high temperature damage. High temperature can denature the bioactive compounds and results in poor retention of the compounds after the drying process and produces a poor-quality product. Often, a temperature that is above 60°C is considered high for this type of material. This poses a limit on the process conditions that we can apply to drying. As such, selection of appropriate drying technology becomes a key aspect in successful processing of the wet material.

Certain particulate solids are hazardous, e.g., combustible or toxic. If the material is combustible, it poses a fire and explosion hazard to the drying operation, especially if the drying is carried out at a high temperature. Organic particulate solids typically have a self-ignition temperature in the range of 200°C–220°C; hence, one will also need to pay attention to the operating drying temperature when dealing with drying organic particulate solids. One way to solve this issue is to inertize the drying chamber in order to remove oxygen from the equation to the extent feasible. If the material is toxic, handling it may require an additional layer of protection to ensure safe operation. If the materials contain toxic solvent, removal of the toxic solvent by drying may require proper containment so that the toxic solvent is captured, condensed, and contained.

Some of the particulate solids are fragile, and they tend to break into smaller pieces when they are subjected to rigorous handling and processing. This includes the handling of friable particulate solids during drying due to the breakage that occurs when particles collide with each other or when they collide with the wall of the dryer chamber. Breakage of friable particulate solids further causes the formation of fine powder and poses a dust explosion hazard.

Many bio-origin products (which contain polysaccharides, sugars, proteins, etc.) are predominantly amorphous materials that can exhibit glass transition. Glass transition temperature is the threshold that indicates whether an amorphous material is in a glassy state or a rubbery state. If drying is carried out at a temperature that is above the glass transition temperature, the amorphous material shifts from a solid and brittle glassy state to a rubbery and viscoelastic state. Dehydrated bio-origin products obtained from most of the common drying processes appear in a glassy amorphous form. At this state, the solid matrix is stable and its mobility is rather limited. Therefore, to ensure the stability of the dehydrated product during storage, this physical state should not alter with time. If the dehydrated material is in a rubbery state, its molecular mobility of the solid matrix is accelerated. This results in changes in terms of physical as well as chemical state, such as sticking, microstructure collapse, agglomeration and caking, loss of volatiles, colour change, and oxidation. These changes affect the product quality and therefore it is advisable to store the dehydrated product at a storage temperature that is below the glass transition temperature. Among the changes mentioned above, stickiness is one of the most important aspects that one has to pay attention to during drying or storage, as it may lead to agglomeration of solid particles and adhesion to the drying chamber wall, also known as wall deposition.

Drying is an energy-intensive unit operation; therefore, its operating cost is high compared with other unit operations in the processing line. First, selection of the appropriate type of drying technology is key to the process industry. This is to ensure the unit operation can produce the desirable product quality. Second, the selected choice requires a minimum of investment. Third, its operating and maintenance costs are within the acceptable range, and lastly it gives the lowest risk of process safety issues, especially in terms of fire or explosion hazards and environmental impact hazards.

1.4 CLASSIFICATION AND SELECTION OF DRYERS

There are a large number of dryer types (over 500) proposed in technical literature, although only about 50 types are commonly used and/or are readily available from various vendors (Mujumdar, 2009; Mujumdar, 2011; Mujumdar, 2014; Lee, Jangam and Mujumdar, 2013). It is impossible to find two dryers that are identical even when used for drying the same material. Even minor changes in feed condition and/or product specification may make the two dryers different in design or in operation or both. Mujumdar (2014), among many others, have provided detailed classification schemes and selection criteria for dryers. For particulate drying, it is important to classify the particles based on their size. In this section, our focus is on providing a brief overview of the more common drying equipment. It is also not extensive enough to cover all types and sub-types of dryers.

There are several book chapters dedicated to classification and selection of dryers in general and specifically for particulate drying (Mujumdar, 2011; Lee et al., 2013; Mujumdar, 2014; Tsotsas and Mujumdar, 2014). It is necessary to classify dryers based on key criteria before we discuss selection strategies. Figure 1.1 shows one such basic classification of dryers for particulate materials based on the mode of operation. Batch dryers are favoured by low throughput, long residence time, batch equipment upstream and downstream, and requirement for batch integrity. However, continuous dryers are favoured in opposite scenarios.

Figure 1.1 shows a detailed classification based on various criteria. We have provided only a few examples of dryers for each category (although there are many more options available). However, the classification shown in Figure 1.1 is rather coarse. Some of the dryers can be further classified into various sub-types.

Baker (1997) has presented an iterative "structural approach" for dryer selection. It includes the following key steps:

- List all key process specifications
- Carry out preliminary selection
- Carry out bench scale tests including quality tests

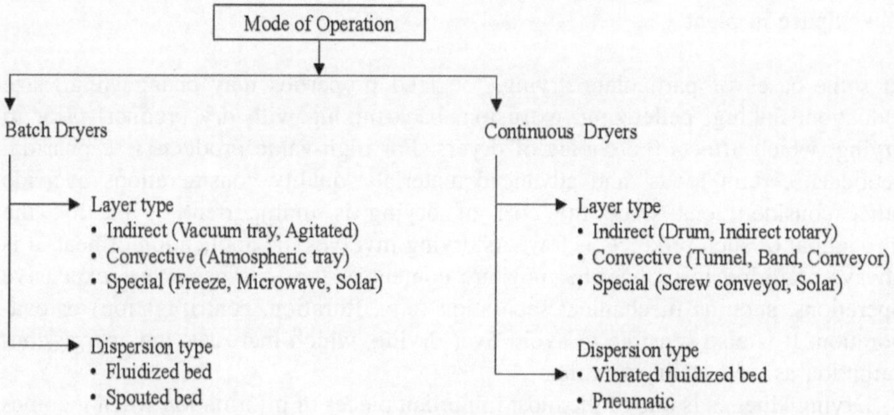

FIGURE 1.1 Basic classification of dryers based on mode of operation.

- Make economic evaluation of alternatives
- Conduct pilot-scale trials
- Select most appropriate dryer types

At times, the selection is solely based on previous experience, however. This approach may have few limitations, especially if the original choice was not optimum. Selection of an appropriate dryer or drying system is a very crucial step. Even if someone has the best design of a wrong dryer, it is not at all useful. Traditionally, dryer selection was made based on experts' experience and knowledge. However, use of more sophisticated computer-based tools in recent times has been helpful. Amongst these possibilities, fuzzy expert systems are the most promising (Baker, 2001). Some of the authors of this chapter have also been working on an online tool for basic selection of dryers.

For selection of a suitable dryer for a given application, one needs at least the following quantitative information:

- Throughput required
- Type of operation needed (batch/continuous)
- Properties of wet feedstock (physical, chemical, and biochemical properties, along with information on possible variations)
- Desired qualities of the final product specifications
- Upstream and downstream processing operations
- Moisture content of the feed and product
- Information on drying kinetics
- Safety aspects, e.g., any possibility of fire hazard and explosion hazards, toxicity
- Value of the product
- Need for automatic control
- Type and cost of fuel available, cost of electricity
- Environmental regulations
- Space in plant

In some cases of particulate drying, the feed properties may change (e.g., size reduction, flaking, pelletizing, extrusion, back-mixing with dry product) prior to drying, which affects the choice of dryers. For high-value products like pharmaceuticals, certain foods, and advanced materials, quality considerations override other considerations since the cost of drying is unimportant. Typically, the throughput of such products is low. As drying involves the application of heat, it is always preferred to reduce the moisture content in the feed using less expensive operations, such as mechanical separation (e.g., filtration, centrifugation) or evaporation. It is also sensible to avoid over-drying, which increases the energy consumption as well as drying time.

Drying kinetics is one of the most important pieces of information when it comes to selection of dryers. It not only helps us decide the residence time, but also decides the types of suitable dryers. The nature of the moisture, location of the moisture, mechanisms of moisture transfer, physical properties of product, and conditions of the drying medium all have importance in the selection of a suitable dryer and the apt operating conditions.

1.5 TRADITIONAL DRYERS USED FOR PARTICULATE DRYING

By far, the most common dryer for small tonnage products is a batch tray dryer. It consists of a stack of trays, or several stacks of trays, placed in a large, insulated chamber in which a hot drying medium is circulated with appropriately designed fans and guide vanes. Often, a part of the exhausted air is recirculated with a fan located within or outside the drying chamber. These dryers require large amounts of labour to load and unload the product. The key to successful operation is the uniform air-flow distribution over the trays as the slowest drying tray decides the residence time required and, hence, dryer capacity. A continuous version of a batch tray dryer, known as a turbo tray dryer, has been successfully used (Mujumdar, 2011; Mujumdar, 2014).

A fluidized bed dryer is another often-used dryer for particulates. Hot, drying gas at a higher pressure is passed through a perforated bed of moist, solid particles. The objective is to supply gas at a velocity greater than the settling velocity of the particles in the fluidization process. The moist solids are raised from the bottom of the drying chamber and suspended in the drying gas. The hot gas around the solid particles reduces the moisture content to a desired level. The drying gas carries the vaporized liquid away and leaves the drying unit. This drying gas may be recycled to improve energy efficiency of the dryer. There are various choices of fluidized bed dryers available for particulates. For most of the pharmaceutical and food applications, a batch fluid bed dryer is an obvious choice because of the comparatively small quantity of wet material to be processed. On the other hand, a batch fluid bed does not work well for drying specialty chemicals and minerals, where a huge throughput is handled. For such applications, a vibrated bed or plug flow fluid bed is used. Recent developments in FBD include mechanically agitated FBD, use of pulsating flow, or use of immersed tubes for efficient heat transfer (Lee, Jangam and Mujumdar, 2013; Mujumdar, 2014).

A cascading rotary dryer is another traditional industrial dryer used for drying particulates. The dryer consists of a rotating cylindrical shell (called a drum), a driving mechanism, and a support structure. The material moves through the drum as it is inclined slightly, with the discharge end at a lower level than the feed end. The wet feedstock enters the dryer and, as the dryer rotates, the material is lifted up by a series of flights. When the material gets high enough, it falls back down to the bottom of the dryer, passing through the hot gas stream as it falls. To increase the retention time of very fine and light materials in the dryer (e.g., cheese granules), in rare cases, it may be advantageous to incline the cylinder with the product end at a higher elevation. There are several other traditional and recent dryers used for particulate matters. This book focuses on some of the conventional and recent drying techniques (Table 1.2).

TABLE 1.2
Classification of Dryers (Mujumdar, 2011; Mujumdar, 2014)

Criterion	Classification
Mode of Operation	• Batch • Continuous
Heat Input	• Convection • Conduction • Radiation • Intermittent/continuous • Combination
Drying Medium	• Air (most common) • Inert gas (CO_2, N_2, etc.) • Superheated steam • Flue gasses
Operating Pressure	• Atmospheric • Vacuum • High pressure • Variable (swell drying)
Drying Temperature	• Above boiling point (superheated steam drying) • Below boiling point (conventional dryers) • Below freezing point (freeze drying)
State of Material to Be Dried	• Stationary (tray) • Moving/agitated/dispersed (fluidized bed, agitated fluid bed, vibrated bed)
Number of Stages	• Single (conventional dryers) • Multiple (e.g., fluid bed→vibrated bed; spray → fluid bed)
Residence Time	• Very short (e.g., pneumatic drying, spray drying) • Medium (batch fluidized bed, spouted bed dryer) • Long (freeze drying, tray dryer)

REFERENCES

Baker, C. G. J. (1997): *Dryer Selection. "Industrial Drying of Foods,"* Baker, C. G. J., Ed.; Blackie Academic & Professional: London, pp. 242–271.

Baker, C. G. J. and Lababidi, H. M. S. (2001): Developments in computer-aided dryer selection. *Drying Technology*, Vol. 19, No. 8, pp. 1851–1873.

Lee, D. J., Jangam, S. V., and Mujumdar, A. S. (2013): Some recent advances in drying technologies to produce particulate solids. *KONA Powder and Particle Journal*, Vol. 30, pp. 69–83.

Mujumdar, A. S. (2009): *Drying in Albright's Chemical Engineering Handbook*, Ed. Lyle F. Albright. CRC Press.

Mujumdar, A. S. (2011): *Drying in Kirk-Othmer Encyclopedia of Chemical Technology.* John Wiley & Sons.

Mujumdar, A. S. (2014): *Handbook of Industrial Drying* (4th Edition). CRC Press.

Tsotsas, E. and Mujumdar, A. S. (2014): *Modern Drying Technology (Volumes 1–5).* Wiley-VCH Verlag GmbH & Co. KGaA.

2 Vibrated Bed Drying

Hugo Perazzini and Maisa Tonon Bitti Perazzini
Institute of Natural Resources, Federal University of Itajubá, Itajubá, MG, Brazil

Lucas Meili
Laboratory of Processes, Center of Technology, Federal University of Alagoas, Maceió, AL, Brazil

Fabio Bentes Freire and José Teixeira Freire
Department of Chemical Engineering, Federal University of São Carlos, São Carlos, SP, Brazil

CONTENTS

DOI: 10.1201/9781003207108-2

2.1 INTRODUCTION

Vibrated beds were developed to overcome difficulties in the fluidized bed (FB) technique when dealing with flat, cohesive particles or materials that are heterogeneous in terms of shape, size, and density, which can lead to the unwanted formation of preferential channels, bubbles, and agglomerates. In this context, the vibro-fluidized bed (VFB) improves the movement of particles, thus increasing the heat and mass transfer rates and decreasing the minimum fluidization velocity, which improves the energy performance of the system. It can be said that vibration is widely found in industrial applications. VFBs can be seen in a variety of unit operations such as cooling, heating, classification, separation and particles coating, and finally drying, which has the largest number of applications, mainly in the pharmaceutical, food, chemical, mining, and textile industry. The first studies on VFBs date back to 1938 and, since then, efforts have been focused on understanding transport phenomena and theoretical aspects of vibration, as can be seen in the pioneering works of Pakowski et al. (1984) and Érdesz et al. (1986). Taking as a starting point the established works on VFBs, the objective of this chapter is to show new discoveries found in the literature that impact the efficient operation of the vibrated bed and improve mathematical description for process design and optimization. The chapter also shows examples of current industrial applications of this type of bed.

2.2 GENERALITIES OF VIBRATED BEDS

2.2.1 The Universality of the Dimensionless Vibration Number

The dimensionless vibration number (Γ) provides a measure of not only the intensity and vigor of the vibration but also the fluid dynamics characteristic of these types of beds. This dimension-less number is given by Equation 2.1, which describes the ratio of vibrational acceleration and gravity.

$$\Gamma = \frac{A(2\pi f)^2}{g} = \frac{Aw^2}{g} \tag{2.1}$$

In recent years, this dimension-less number has been analyzed in the literature mainly because some authors believe that it is a universal parameter. However, there are studies that show that the same Γ can have different combinations of A and f, leading to different operating conditions. Daleffe and Freire (2004) and Meilli et al. (2012a) showed that the same value of Γ can lead to different fluid dynamic behavior in drying pastes and solutions. The fluid dynamics of a VPB with Geldart C particles strongly depends on different combinations of A and f for constant Γ (Meili et al., 2012b). The same happened in Perazzini et al. (2017), in which the drying kinetics parameters were estimated by fitting the experimental data of drying of alumina particles in a VFB to the Page equation. Different values of the drying kinetics constant k and the parameter n of the Page model resulted in the same Γ, even for the same surface air velocity and temperature. The effective moisture diffusivity was likewise different for the same Γ in Meili et al. (2020). Zhao et al. (2015) analyzed the segregation of particles, showing that there are different behaviors for a given Γ with different A and f. In the above-mentioned examples, it is clear that a given Γ can lead to different drying behavior as it is a parameter that allows for different combinations of A and f, as given by Equation 2.1. Therefore, both fluid dynamics and heat and mass transfer must be analyzed for different combinations of A and f, since Γ is not a universal parameter.

2.2.2 Fluid Dynamics

Fluid dynamics of a FB have a clear and well-defined region in the transition from the fixed bed to the FB. In general, vibration causes the characteristic curve to be dampened and there is no maximum pressure point. Compared to the FB, the dynamic behavior of the bed depends on the vibration. The ΔP after the beginning of fluidization is lower than in the FB, and even for a given value of Γ, the fluid dynamics strongly depends on A and f, individually. The vibration can either increase or decrease the minimum fluidization velocity (u_{mf}), depending on the combination of A and f. Without vibration, the transition from the fixed bed to the fluidization condition is easily noticeable. However, with vibration, this transition can be within a considerable range of surface air velocity (u). This encouraged the development of correlations to determine the minimum fluidization velocity for VFBs (u_{mfv}). Some of these correlations show that increasing Γ in constant A decreases the ΔP. The increase in Γ due to the increase in the vibration frequency improves the air passage through the bed. Furthermore, bed expansion leads to a delay in the onset of fluidization. Vibration led to a reduction in ΔP in the region of minimum fluidization velocity of around 25% in relation to the bed without vibration. Thus, it can be said that vibration is of great importance when it is desired to reduce ΔP. From Equation 2.2, it is clear that the magnitude of u_{mfv} depends on the relation of the pressure drops at different minimum conditions (Érdesz et al., 1986).

$$u_{\mathrm{mfv}} = 0.8u_{\mathrm{mf}}\left(\frac{\Delta P_{\mathrm{umf}}}{\Delta P_{\mathrm{umfv}}}\right)^2 \tag{2.2}$$

2.2.3 Heat and Mass Transfer

Heat transfer in vibrating beds occurs mainly by gas-particle energy convection. The extent of these contributions in quantifying the value of the heat transfer coefficient (h) depends on the vibrational acceleration range, the gas velocity, and also the bed properties. Picado and Martínez (2012) found differences of up to four orders of magnitude between the values of h of a total of nine different correlations. Some of these correlations are only function of Γ, as the correlation of Rezende and Finzer (2008):

$$Nu = \frac{hd_p}{\lambda_g} = 0.1926Re^{0.156}\Gamma^{0.889} \tag{2.3}$$

The pioneering work by Pakowski et al. (1984) shows that heat transfer is more the greater the bed vibration, but only up to a given value of Γ. From $\Gamma = 6$ on, h starts to decrease, showing that the bed reached an operating condition in which the vibration acceleration was maximum, increasing the thermal energy transfer in the system.

Contributions to a better understanding of mass transfer are rare in the literature. Thus, a few studies were found on correlations for the mass transfer coefficient. Borde et al. (1997) predicted the mass transfer in grain drying with the following correlation:

$$Sh = \frac{d_p}{D_v}\left(\frac{k_y}{\rho_g}\right) = \frac{2 + 0.6Re^{0.5}Sc^{0.333}}{\left[1 + \frac{c_{pw}\left(T_g - T_s\right)}{\Delta H_v}\right]^{0.7}} \tag{2.4}$$

2.2.4 Energy Performance

VFB differs from FB in that it improves the motion of sticky particles and thus increases heat and mass transfer rates, leading to better energy efficiency. In fact, Meili et al. (2020) found the energy spent of the VFB to be 5,000 kJ kg^{-1}, within the range shown by Mujumdar (2015) (4,000–6,000 kJ kg^{-1}) for FBs. Perazzini et al. (2020) showed that the energy efficiency of the VFB was in the range of 10–48%, which is compatible with the efficiency levels Strumillo et al. (2015) for FB.

2.2.5 Factors Influencing Process with Vibration

For the process to be efficient, it is essential to find the key vibration factors that improve bed operation. Spouting, for example, may occur during vibration. The technique causes the solids to have a circular movement up through a central channel and down through the sides (Rátkai, 1986). The addition of vibration helps the initial spouting by improving the breakdown of the bed of particles. On the other hand, vibration can lead to preferential air channels in the case of thick beds. This may be due to the distribution of air in the bed both by the distributor and by the bed itself; in addition, there may be dead zones in the corners of beds with rectangular geometry. From an experimental point of view, this happens when there are sudden pressure fluctuations, that is, when ΔP reaches a maximum value and then abruptly decreases.

The dispersion of particles is a form of mixing that involves the diffusion of molecules at microscopic levels in either horizontal (Dong et al., 2015) or vertically vibrated beds with continuous flow (Finzer et al., 2007). It strongly depends on the concentration of particles in the bed and also on the combination of vibration amplitude and frequency (Dong et al., 2015). Dispersion has to do with the distribution of the residence time of the particles in the bed and can be evaluated from Taylor or Free Dispersion models (Finzer et al., 2007). Finzer et al. (2007) found that if the Peclet number (Pe) is in the magnitude of ε_b, molecular diffusion is predominant; if Pe is in the order of one, the diffusion is on a small scale; if Pe is in the order of Pe^{-1}, convection tends to be significant; the convection is more significant when Pe is of the order of Pe^{-2} (Finzer et al., 2007). Dong et al. (2015) found that the low concentration of particles in the bed leads to a more uniform dispersion and less oriented particles, while the high concentration of particles in the bed results in greater energy dissipation between the particles.

Particle segregation is a complex phenomenon in particulate systems with large particles together with smaller granular materials within a wide range of density, friction, and restitution coefficient, which can be reduced by vibration and external drying conditions (Liao and Hsiau, 2016). It is a key parameter in vibrated bed applications to classify particles in the chemical or coal industries. Segregation of particles in vibrated beds has been extensively studied. Zhao et al. (2015) showed that by increasing A and decreasing f, the ash separation increased significantly, as there were more expanded beds, improving the percolation of air through the bed. Cano-Pleite et al. (2017) evaluated particle segregation with time with the modified Lacey Mixing Index (SI). It was found that the SI increased with the increase in the surface velocity of the gas, but decreased with the increase in the dimension-less vibratory number, showing that the high-vibration intensity led to more convective bed movements and more particle mixing, thus neutralizing the segregation effect of the gas velocity.

The agglomeration, in its turn, is due to cohesion of the particles. Cohesive forces play an important role in the movement of wet particulate solids as in this case there may be surface adhesion, van der Walls attraction forces or interstitial forces arising from liquid bridges (Hsiau et al., 2004). Under specific vibration conditions, cohesive forces can decrease to improve solids circulation in the bed and also decrease the minimum fluidization velocity. Hsiau et al. (2004) showed that high f together with low Γ (between 1.8 and 2) decreased the value of u_{mf} due to the increase in vibration intensity. There are several models in the literature to predict the phenomenon of agglomeration in fluidization (Raganati et al., 2018). In engineering practice, it is interesting to estimate the agglomeration degree of a given system. Although no works were found on this subject, it is believed that prediction through models can be adequately performed under moderate vibration conditions. Jaraiz et al. (1992) proposed a dimensionless measure of relative significance of the interparticle forces in vibrated beds (R_c). If $R_c > 1$, then interparticle forces are significant due to the cohesiveness between the particles and if $R_c < 1$, then the gravity and inertial effects are significant in fluidization.

While there are ways to decrease the size of both preferred air channels and bubbles in vibrated beds, depending on particle size and density and surface air velocity, it is hard to avoid them altogether. Cano-Pleite et al. (2013) showed that vibration increased the size of bubbles in the fluidization of 200 μm diameter ballotini glass spheres. There were no bubbles at surface gas velocities close to the minimum fluidization conditions, but for a higher surface gas velocity, vibration not only increased the size but also the growth of the bubbles with distance from the distributor in relation to the non-vibrating beds. Zhou et al. (2021) showed that if, on the one hand, the bubble size increased with the increase in the height of the bed of fine particles of coal, on the other hand, the number of bubbles decreased. The authors found that the greater the amplitude and frequency of vibration, the smaller the sizes and quantities of bubbles, with a decrease of almost 50% for $u_{mf} = 0.16$ m/s, $A = 2$ mm and $f = 20$ Hz. Zhou et al. (2021) verified that the bubble diameter depends on the porosity of the bed at minimum conditions, the density of the particles and the standard fluctuation of the bed pressure drop.

2.3 MATHEMATICAL MODELING FOR PROCESS OPTIMIZATION AND DESIGN

The mathematical models can be divided in three main groups: single-particle approach, single-phase model, and two-phase model.

One of the available models for the single-particle approach is the two-compartment model (Claudio et al., 2022). In this model the hypothetically spherical particle has two well-defined compartments: an inner part without direct contact with the air stream, in which moisture diffusion is the primary transport mechanism, and an outer layer in which there is both the diffusion of the first compartment and the convection of vapor on the surface in contact with the drying air. In terms of the moisture ratio ($MR = X - X_{eq}/X_0 - X_{eq}$), the following set of differential equations can be written for each compartment:

$$\frac{dMR_1}{dt} = -k_1(MR_1 - MR_2) \tag{2.5}$$

$$\frac{dMR_2}{dt} = -k_1(MR_2 - MR_1) - k_2 MR_2 \tag{2.6}$$

The average values of MR as a function of time should be determined accordingly:

$$\overline{MR} = MR_1\tau_1 + MR_2\tau_2 \tag{2.7}$$

Parameter τ is a mathematical discretization ($\tau_1 = \tau_2 = 0.5$, for simplicity) and k_1 and k_2 are estimated as a function of the air temperature following an Arrhenius relationship.

In the single-phase model, the particles are assumed to be perfectly mixed in the vibrated bed such that an overall heat balance can be made. Taking the spherical particle as the control volume, the temperature variation of the single particle can be written as:

$$\rho_s c_{ps}\frac{dT}{dt} = ha_v(T_g - T) - \rho_s\left(-\frac{d\overline{X}}{dt}\right)\Delta H_v \tag{2.8}$$

Under thin-layer condition, a moisture balance equation can be derived from the differential form of the diffusive model with the convective-type boundary condition (Nagata et al., 2020):

$$\frac{\partial \overline{X}}{\partial t} = -(X_0 - X_{eq})6Bi_m^2 \sum_{n=1}^{\infty} \frac{1}{\left[\gamma_n^2 + Bi_m(Bi_m - 1)\right]}\left(\frac{D_{eff}}{R^2}\right)\exp\left[-\lambda_n^2\left(\frac{D_{eff}}{R^2}t\right)\right] \tag{2.9}$$

The coupling between Equations 2.8 and 2.9 is done by an Arrhenius-like relationship with D_{eff} as a function of the solid temperature.

Mass and energy balances for the solid and fluid phases can be performed when the VFB is assumed to be a perfect mix reactor. The resulting differential equations from heat and mass balances for the fluid and solid phases are given by Equations 2.10–2.13.

$$\frac{dX}{dt} = -k(X - X_{eq}) \tag{2.10}$$

$$\frac{dY}{dt} = \frac{\dot{V}}{M_g}\rho_{g,i}(Y_i - Y_g) + \left(\frac{M_s}{M_g}\right)k(X - X_{eq}) \tag{2.11}$$

$$\frac{dT_s}{dt} = \frac{1}{c_{ps} + Xc_{pw}}\left[\frac{hA(T_g - T_s)}{M_s} - k(X - X_{eq})\Delta H_v\right] \tag{2.12}$$

$$\frac{dT_g}{dt} = \frac{\dot{V}}{M_g}\rho_{g,i}(T_{g,i} - T_g) + \frac{1}{c_{pg+}Yc_{pv}}\left[-\frac{hA(T_g - T_s)}{M_g} - \left(\frac{M_s}{M_g}\right)k(X - X_{eq})\Delta H_v - \left(\frac{M_s}{M_g}\right)k(X - X_{eq})c_{pv}(T_g - T_s) - Q_1\right] \tag{2.13}$$

This set of differential equations describes the change of moisture and temperature of the solid and the humidity and temperature of the gas over time. Variables like X_{eq}, k, h, M_s, M_g, and Q_1 can be obtained from specific studies or correlations proposed by literature.

2.4 TYPES OF DRYERS WITH VIBRATION

2.4.1 VIBRATION-GENERATING DEVICES

Vibration is provided to the particulate system through specific devices. The most common mechanisms found in applications are the electromechanical vibrator, the electromagnetic vibrator, and the pneumatic vibrator. The electromechanical vibrator, also known as a motor vibrator, generates a rotational force with an eccentric shaft that makes the dryer vibrate. The electromagnetic vibrators have a coil and a core (inductor) that acts as a magnet from the moment it is energized. When the magnetic flux increases, the armature is attracted by the inductor and when the flux decreases, it retreats by the action of an elastic system. Unlike electromechanical vibrators, these electromagnetic vibrators generate a more linear vibration. This device is robust, able to work for several years with very little maintenance, and widely found in applications with a high frequency of vibration. Pneumatic vibrators, in turn, require a constant flow of air to move a piston. They provide rotational motions, with high frequency of vibration and low energy consumption, as well as linear motions.

2.4.2 Description of Different Types of Dryers with Vibration

2.4.2.1 Horizontal-Vibrating Fluidized Bed

On an industrial scale, the most common vibrating dryers are continuous ones with beds positioned horizontally, or with a slight inclination (Figure 2.1). These dryers are of the plug-flow type, in which the solid is fed at one end, usually by screw conveyors or conveyor belts, depending on the physical properties of the particulate material. The solid, in turn, is placed on a screen that allows the passage of fluid going from the bottom to the exhaust at the top of the dryer. The solids move through the drying chamber and are removed at the other end of the dryer. In this type of dryer, vibration is provided by metal springs and motor vibrators.

The horizontal-vibrating fluidized bed can be found in many industrial applications, mainly in the minerals and polymers industry, where they both dry and cool solids, and also in the drying of materials that are difficult to fluidize, such as tea, leaves, and biofuels. They can also be found in the chemical and pharmaceutical industries and in milk drying as a second-stage dryer to improve product quality while maintaining the organoleptic properties of milk. The superficial area is in the range 0.5–10 m^2, 20–250 kg h^{-1} evaporation rate, and 0.75 – 3 × 2 kW vibration motor power.

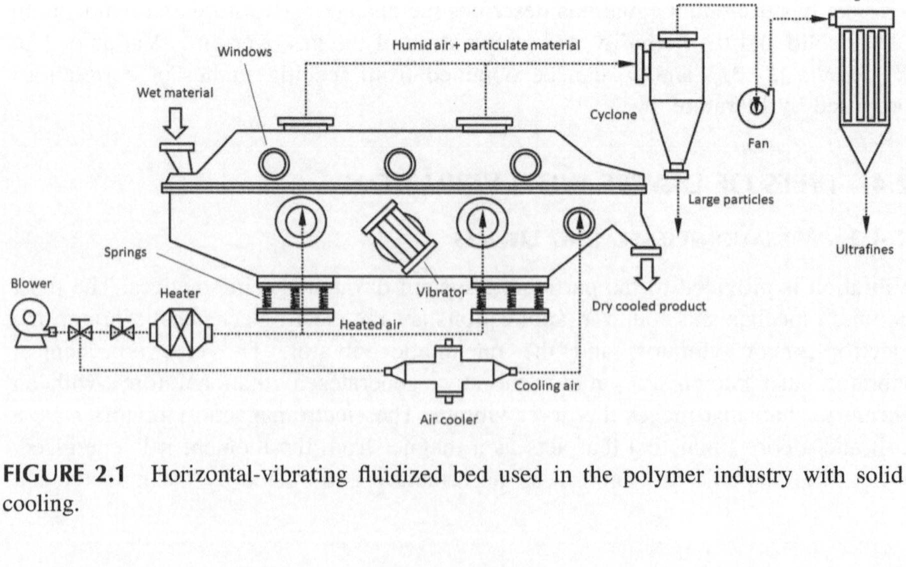

FIGURE 2.1 Horizontal-vibrating fluidized bed used in the polymer industry with solid cooling.

2.4.2.2 Vertical-Vibrating Fluidized Bed

Vertical-vibrating dryers are usually made with multiple trays, as is the vibrating tray dryer shown in Finzer et al. (2007) (Figure 2.2a), or spiral form, as in the work of Erdesz et al. (1989) (Figure 2.2b). Both are found in grain drying, with a wide range of residence times, in addition to several other industrial applications, such as polymers, pharmaceuticals, food, and chemical products. They normally have four main parts: a vibration system, a vertical drying tunnel, heating, and air transport system, and occasionally a solids recycling system. The wet solid is fed at the top and is arranged in perforated trays that allow upward airflow. The trays are slightly angled to improve the solids' flowability in the drying chamber. The vibration system usually consists of electromagnetic vibrators connected to each tray. The vibration of the entire drying chamber is not recommended, as it increases energy costs and reduces the solids' circulation rate. In vertical VFB with circular trays, particulate material can enter from the bottom or from the top, and the vibrator is installed in the opposite position to the solids' feed. The advantage of this dryer is that it requires less area in the industrial plant than a horizontal dryer and also provides vertical transport of particles, either downward or upward, to tanks, hoppers, or other unit operations after drying.

It is noteworthy that the transport of particles strongly depends on the combination of amplitude and frequency of vibration, since excess vibration further deteriorates the helical transport of solids and leads to non-homogeneous drying. These dryers generally do not have the same productivity levels as horizontal dryers. A heating jacket ensures uniform heating in the drying chamber as solids are transported. Vibration motor power are usually in the range $1.1 - 2.2 \times 2$ kW, $3.9 - 13.2$ m^3 overall dimension, and $500 - 4{,}000$ kg h^{-1} nominal capacity.

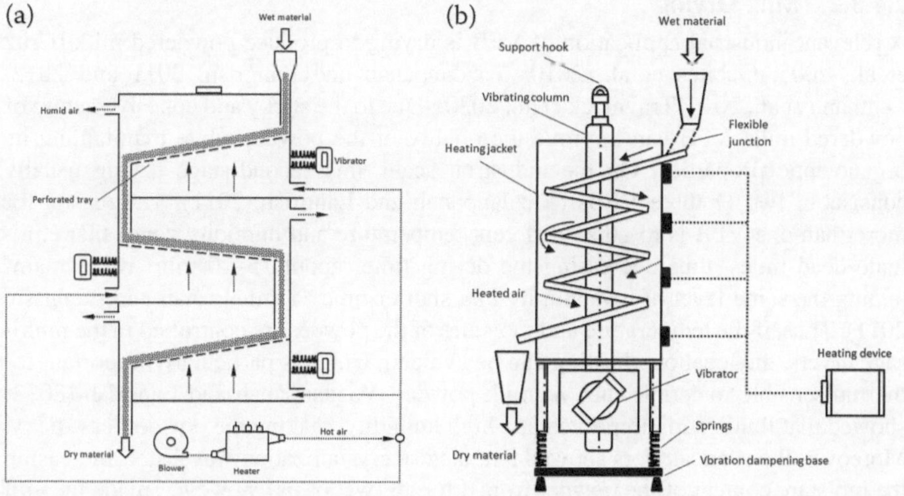

FIGURE 2.2 Vertical-vibrating fluidized bed for particulate material drying: (a) with multiple trays; (b) with spirals.

2.4.3 Applications

2.4.3.1 Particulate Inorganic Compounds

The fluidization of particulate inorganic compounds is generally heavy and both the pressure drop in the bed and the energy consumption are high. In fact, Perazzini et al. (2020) showed that the drying of alumina particles in VFB was more efficient than in FB. Only after the free water was completely removed did the bed work as expected. The fluidization efficiency increased after the superficial water was removed. In this context, the main role of vibration is to improve the homogenization of particles in the bed. High levels of energy efficiency could be achieved even by increasing the u_{mf}, as long as the optimum combination of amplitude and frequency is found, i.e., low amplitude and higher frequency, with $\Gamma = 4$ ($A = 0.003$ m $f = 18.20$ Hz). This would be a way to reduce expenses. Similarly, Lehmann et al. (2020) showed that vibration decreased the agglutinating effect of the cohesive forces between the wet particles, and also decreasing the u_{mf} of the alumina beads. It is known that FB is found in the polymer industry; however, after the polymerization step, the granular polymer particles can be very sticky and wet, leading to preferential air paths with bubbles. This, in turn, can lead to elutriation of the particles and also to excessive drying of the product, making visible black spots on the polymer surface if colored pigments are used. In this sense, although rarely found in literature, VFBs are an alternative to drying polymers to ensure the quality of the final product. VFBs can be applied in the polymer industry to dry, cool, and transport particulate material. That is to say, this can be done in a single setup. According to Mujumdar and Hasan (2015), the VFBs work with a frequency of 5–25 Hz and amplitude of 0.002–0.005 m.

2.4.3.2 Milk Drying

A relevant industrial application of VFB is drying to produce powdered milk (Cruz et al., 2005; Gabites et al., 2010; Yazdanpanah and Langrish, 2011 and 2012; Lehmann et al., 2019; Trávnícek et al., 2020). Due to the sticky and cohesive nature of powdered milk, VFB can improve the quality of the powder while maintaining its organoleptic properties. On the industrial scale, this second-stage drying usually consists of two or three VFBs (Yazdanpanah and Langrish, 2011). The reason for more than one VFB is to allow different temperature and humidity zones that eliminate dead times, thus decreasing the drying time (about 15–20 min) while maintaining the same level of crystallinity in a shorter time (Yazdanpanah and Langrish, 2011). Thus, if the temperature and moisture of the powder are controlled in the multisage dryers, the local overheating can be avoided, which is particularly important for thermo-sensible materials such as milk powder. Yazdanpanah and Langrish (2012) showed that fluidization improved at a high humidity, making the powder less sticky. Moreover, the same authors showed that lactose crystallization was faster. Increasing the moisture content of the powder from 4 to 9% (wet basis), however, made the umf increase by a factor of six, but the vibration of intensity 1.21 caused the minimum fluidization velocity to decrease by 50%, with a decrease in bed expansion and bubble volume fraction (Lehmann et al., 2019). The contact between the particles and agglomerates was improved by the increase of the vibration amplitude, but increasing the air velocity from $1.2u_{mf}$ to $2.5u_{mf}$ increased the elutriation of fine particles (Cruz et al., 2005).

2.4.3.3 Leaves, Tea, and Herbal Medicines

Flat particles, with low sphericity and specific mass, such as herbs, tea, and leaves in general, are hard to fluidize conventionally. Due to the peculiar morphology, such as irregular shape, size heterogeneity, and random geometry, the particles strongly resist fluid flow, reducing air percolation and leading to preferential channels that increase pressure drop and consequently drying costs. According to Lima and Ferreira (2009), particles of low sphericity can lead to different fluid dynamic characteristics when compared to FB with regular particles. The authors showed that the most likely spatial orientation for a bed of flat particles is entirely random, rather than parallel or perpendicular to the direction of air flow. Thus, VFB improves not only the fluidization of flat particles but also the contact between the fluid and the solid phase.

In the study on fluidization and vibrofluidization of fresh leaves in shallow beds, Lima and Ferreira (2011) found that continuous vibration improves fluidization, leading to orderly arrangement of leaves while increasing heat and mass transfer rates due to a better contact. In VFB, the combination of floating and mixing avoided preferential channels and also the agglomeration of particles, improving the gas-solid contact. In contrast to FB, VFP allowed for more uniform coloring and moisture without leaf agglomerates, proving to be, under these operating conditions, a perfect mixing tank. Thanimkarn et al. (2019) dried *Cissus quadrangularis*, a medicinal plant found in Thailand, in an infrared-vibrated fluidized bed (VFBI) and also in a vacuum-vibrated infrared bed (VFBVI). The combination of infrared and vacuum allowed to increase the drying rate even with poor contact between the particles and the drying air, common in this type of material, as shown in previously mentioned works. Although the heat and mass transfer rates of the VFBVI reduced drying time, the drying rates were not high enough to achieve high energy efficiency. The vacuum also ended up decreasing the total phenolic content and the quercetin content of the dry material. When drying green tea, Handayani et al. (2017) showed that the decrease in drying time strongly depends on both the velocity and temperature of the inlet air. In this work, the temperature in the drying chamber was homogeneous, as the temperature at the dryer exhaust increased during drying. Drying was done in the conventional continuous VFB, where the fermented leaves are placed on a vibrated grid. According to Chen and Mujumdar (2015), VFB is widely found in tea leaf drying, as it improves the fluidization of wet particles and reduces the amount of fine particles, thus improving the quality of the final product.

2.4.3.4 Biomass and Biofuels

Biomass particles are irregular in shape and size, and together with the low bulk density, the fluidization is quite challenging, requiring the use of inert materials or special air distributors to improve bed hydrodynamics. The vibration added to the extra acceleration provided by the pulsation of the gas can overcome the difficulties found in the fluidization of cohesive particles. Vibrated dryers are found in combined heat and power plants, in which dry biomass is transferred to fluidized bed reactors to extract and release product gases from biomass materials (Mohseni et al., 2019). Due to the characteristics of biomass fluidization, changes must be made either in the design or operational mode of the conventional VFB. Jia et al. (2015) dried biomass such as pine, switchgrass, and Douglas fir in a pulsed fluidized bed with vibration (PFVB). Pulsation frequency improved particle circulation and, along with vibration, helped to break down interparticle forces. According to Jia et al. (2015), gas-solid interactions improved with pulsation because of the intermittent acceleration and deceleration of particle motion. Zhao et al. (2016) dried Shengli lignite in a fluidized bed, conventional vibrated bed, assisted medium fluidized bed, and vibrated medium-assisted fluidized bed (VMFD). The comparison between them showed that VMFD had better performance in both drying and fluidization, with higher rates and shorter drying times. In this case, mechanical vibration together with the lubricating action of inert particles led to fluidization of hard-to-fluidize particles.

A magnetic medium, such as magnetite, has a high thermal conductivity and can increase the drying rate and improve fluidization of wet materials with less air than conventional techniques. This decreases both energy consumption and particle elutriation (Zhao et al., 2016). Thus, the removal and recovery of wet biomass magnetite can be done through an external magnetic field. Zhao et al. (2016) showed that drying lignite in this type of vibrated bed can significantly reduce process costs. Conventional vibrated beds have been widely found in separation processes with great potential for the coal processing industry (He et al., 2015; Mujumdar et al., 2015). In addition, they are also found in the processing of brown and hard coal with particle sizes in the range of 3–32 mm, including drying and solids transport (Mujumdar et al., 2015), and for the desulfurization of fine coal with significant reduction of SO_2 during combustion, including efficient removal of sulfate and pyrites (Dong et al., 2015). For 1–6 mm sized coal particles, there was particle segregation (He et al., 2015). On the other hand, dry separation of fine carbon particles (below 6 mm in diameter) can be done in a VMFD with a dense medium of magnetic particles (size between 0.074 mm and 0.3 mm) within the Geldart B group (Zhou et al., 2017). In this case, the vibrated dense medium fluidized bed (VDMFB) requires less drying air and is therefore more energy efficient. The diffusion of particles in the VDMFB strongly depends on physical properties such as density, diameter, and sphericity. Zhou et al. (2017) showed that VDMFB was efficient to separate fine coal by the dry method, with a recovery rate of combustible components in the order of 87.33%. The main bottleneck of vibrated beds in dry coal separation is that segregation performance and separation quality are strongly dependent on operating conditions, such as surface gas velocity, vibration intensity, bed height, and time of fluidization (Yang et al., 2013). As the bed height increases, bubble motion plays an important role due not only to the increase in bubble sizes but also to the decrease in the number of bubbles (Zhou et al., 2021). Yu et al. (2018) showed that there were bubbles in the drying of the oil shale in VDMFB with ferrosilicon powder as the medium. In this study, the formation of bubbles was little dependent on vibration. The bed density distribution was more uniform with vibration, preventing material settlement. Low amplitude, high-frequency vibration improved drying for oil shale purification.

2.4.3.5 Agroindustrial Residues

Recently, VFB has been found in the thermal treatment of agro-industrial residues for renewable energy generation. When drying spent coffee ground powders, Rocha et al. (2021) showed that good process stability was achieved with 0.015 m amplitude and $\Gamma = 4$, without changing the mean Sauter diameter of the elutriated powder. High air velocities from 4 to 6 u/u_{mf}, ensured a high rate production of the dry powder, with $u/u_{mf} = 6$ giving the optimum moisture content of 22% w.b. and mean particle size of 0.37 mm. Special attention must be paid to the airflow rate as it significantly changes the powder feed rate and size distribution. Vibration increased the powder feed rate. Barbosa et al. (2020) dried orange bagasse under these same vibration conditions. A mechanical dewatering previous step with CaO was performed before drying to reduce the initial moisture content of the residues. There was a decrease of approximately 40% w.b., and the drying kinetics did not depend on CaO.

A summary of the operating conditions works found in literature is shown in Table 2.1.

2.5 FINAL REMARKS

The main characteristics and parameters of vibrating beds were analyzed with an emphasis on applications where a conventional FB is not suitable. Recent and relevant technical-scientific studies were the basis for the discussion of the topic. The main types of VFB dryers and mathematical models for process design, optimization and control were considered. A better background on bed fluid dynamics helps to understand how particle motion and typical VFB phenomena, such as segregation, agglomeration, spouting, and dispersion of particles, depend on vibration. Although heat transfer has been extensively studied, it is necessary to update the topic. Mass transfer, in turn, has been little studied in the context of vibrated beds, with very few works found in literature. The mathematical models that assume piston air flow in the VFB are very complex, with a high computational cost. Simpler models, such as CST or single-particle models, tend to oversimplify the complexities of this topic, while still providing satisfactory estimates. There are many industrial applications of VFB, but above all, some stand out, such as drying heterogeneous matter like biomass and biofuels, in the context of renewable energy and coal processing, in addition to being an interesting alternative to combined power plants of heat and energy due to the ability to classify, transport, and dry particulate material.

TABLE 2.1
Typical Vibrating Conditions of Vibrated Beds

Type	Material	Material Properties	Bed Properties	Vibrating Conditions	Reference
VFB	Alumina	$d_p = 3.68 \times 10^{-3}$ m $\rho = 1750$ kg m^{-3} $X_0 = 0.32$ (d.b.)	$\varepsilon = 0.42$ $H = 0.10$ m	$A = 3 \times 10^{-3}$ m and $f = 18.20$ Hz $A = 0.015$ m and $f = 8.14$ Hz	Perazzini et al. (2020)
VFB	Milk powder	$d_p = 1.02 \times 10^{-4}$ m $\rho = 866$ kg m^{-3} $4 < X_0$ (%, w.b.) < 9	$0.056 < \varepsilon < 0.61$ $H = 3.0$ m	$3 \times 10^{-3} < A$ [m] $< 5 \times 10^{-3}$ $4 < f$ [Hz] < 10	Lehmann et al. (2019)
VFB	Coffee ground powders	$d_p = 3.62 \times 10^{-4}$ m $\rho = 1103$ kg m^{-3} $X_0 = 45\%$ (w.b.)	$\varepsilon = 0.57$	$3 \times 10^{-3} < A$ [m] < 0.015 $\Gamma = 4$	Rocha et al. (2021)
VFB	Citrus residues	$d_p = 0.07$ m $75 < X_0$ (%, w.b.) < 85	$H = 0.043$ m	$A = 0.015$ m and $f = 6.16$ Hz	Barbosa et al. (2020)
PFVD	Woody biomass	$7.55 \times 10^{-4} < d_p$ [m] $< 1.45 \times 10^{-3}$ $1375 < \rho$ [kg m^{-3}] < 1446	$0.17 < H$ [m] < 1.5	$0 < A$ [m] < 0.21 $0.33 < f$ [Hz] < 3	Jia et al. (2015)
VMFD	Shengli lignite and magnetic powder	$1 \times 10^{-3} < d_p$ [m] $< 3 \times 10^{-3}$ $\rho = 1.3$ kg m^{-3} $X_0 = 39.58\%$ (w.b.) $d_p = 232 \times 10^{-6}$ m (magnetic powder) $\rho = 4600$ kg m^{-3} (magnetic powder)	$H = 0.60$ m	$A = 2 \times 10^{-3}$ m and $f = 20$ Hz	Zhao et al. (2016)
VDMFB	Coal and magnetic powder	$1 \times 10^{-3} < d_p$ [m] $< 6 \times 10^{-3}$ $130 < \rho$ [kg m^{-3}] < 200 $7.4 \times 10^{-5} < d_p$ [m] $< 3 \times 10^{-4}$ (magnetic powder) $\rho = 423$ kg m^{-3} (magnetic powder)	$0.01 < H$ [m] < 0.08	$2.3 \times 10^{-3} < A$ [m] $< 3 \times 10^{-3}$ $28 < f$ [Hz] < 30	Zhou et al. (2017)

NOMENCLATURE

a	Specific area	$m^2\ m^{-3}$
A	Amplitude	m
c_p	Specific heat	$kJ\ kg^{-1}\ °C^{-1}$
d	Diameter	m
f	Frequency	Hz
g	Gravitational acceleration	$m^2\ s^{-1}$
h	Heat transfer coefficient	$W\ m^{-2}\ °C^{-1}$
H	Bed height	m
k	Drying kinetics constant	s^{-1}
k_y	Mass transfer coefficient	$kg\ m^{-2}\ s^{-1}$
L	Length	m
M	Mass	kg
P	Pressure	bar
R	Particle radius	m
Q	Heat transfer rate	W
t	Time	s
T	Temperature	°C
u	Gas velocity	$m\ s^{-1}$
$\dot{V}$	Volumetric flow rate	$m^3\ s^{-1}$
w	Angular velocity	rpm
X	Moisture content	$kg\ H_2O\ kg\ dry\ material^{-1}$
Y	Absolute humidity	$kg\ H_2O\ kg\ dry\ air^{-1}$

Greek symbols

γ	Eigenvalues	
ΔH_v	Latent heat of vaporization	$kJ\ kg^{-1}$
ε	Porosity	
λ	Thermal conductivity	$W\ m^{-1}\ °C^{-1}$
ρ	Specific mass	$kg\ m^{-3}$
τ	Volume fraction	

Subscripts

1	First compartment
2	Second compartment
a	Ambient air
b	Bed, bulk
eq	Equilibrium
g	Gas
i	Inlet
l	Loss
p	Particle
s	Solid
umf	Minimum fluidization
umfv	Minimum fluidization for vibration

v	Vapor
w	Water

Dimension-less numbers

Bi	Biot
Re	Reynolds
Sc	Schmidt
Sh	Sherwood
Γ	Dimension-less vibration number

REFERENCES

Barbosa, A. M., T. A. F. Rocha, J. F. Saldarriaga, I. Estiati, F. B. Freire, J. T. Freire. 2020. "Alternative Drying of Orange Bagasse in Vibrofluidized Bed for Use in Combustion." *Chemical Engineering and Processing: Process Intensification* 152: 107941. doi: 10.1016/j.cep.2020.107941

Borde, I. M., Dukhovny, T. Elperin. 1997. Heat and Mass Transfer in a Moving Vibrofluidized Granular Bed. *Powder Handling & Processing* 9: 311–314.

Cano-Pleite, E., J. Gómez-Hernández, J. Sánchez-Prieto, A. Acosta-Iborra. 2013. "Characterization of the Bubble Behavior in Vibrated Fluidized Beds by Means of Two-Fluid CFD Simulations Coupled with Accelerometry Data." Paper presented at The 14th International Conference on Fluidization – From Fundamentals to Products, 1–9.

Cano-Pleite, E., F. Hernández-Jiménez, A. Acosta-Iborra, C. R. Müller. 2017. "Reversal of Gulf Stream Circulation in a Vertically Vibrated Triangular Fluidized Bed." *Powder Technology* 316: 345–356. doi: 10.1016/j.powtec.2017.01.032

Chen, G., A. S. Mujumdar. 2015. "Drying of Herbal Medicines and Tea." in *Handbook of Industrial Drying*, 4th ed., edited by A. S. Mujumdar, 637–646. Boca Raton: CRC Press.

Claudio, C. C., M. T. B. Perazzini, H. Perazzini. 2022. "Modeling and Estimation of Moisture Transport Properties of Drying of Potential Amazon Biomass for Renewable Energy: Application of the Two-compartment Approach and Diffusive Models with Constant or Moisture-Dependent Coefficient." *Renewable Energy* 181: 304–316. doi: 10.1016/j.renene.2021.09.054

Cruz, M. A. A., M. L. Passos, W. R. Ferreira. 2005. "Final Drying of Whole Milk Powder in Vibrated-Fluidized Beds." *Drying Technology* 23: 2021–2037. doi: 10.1080/07373930500210473

Daleffe, R. V., J. T. Freire. 2004. "Analysis of the Fluid-Dynamic Behavior of Fluidized and Vibrofluidized Bed Containing Glycerol." *Brazillian Journal of Chemical Engineering* 21: 33–46. doi: 10.1590/s0104-66322004000100005

Dong, L., Y. Zhang, Y. Zhao, H. Wang, Y. Wang, Z. Luo, H. Jiang, X. Yang, C. Duan, B. Zhang. 2015. "Deash and Desulfurization of Fine Coal Using a Gas-Vibro Fluidized Bed." *Fuel* 155: 55–62. doi: 10.1016/j.fuel.2015.03.073

Érdesz, K., A. S. Mujumdar, D. U. Ringer. 1986. "Hydrodynamic Similarity of Conventional and Vibrated Fluidized Beds." In *Drying '86*, edited by A. S. Mujumdar, 169–176. New York: Hemisphere.

Erdesz, K., A. S., Mujumdar. 1989. "Fundamentals of Drying Techniques". *Drying Technology* 7: 401–410.

Finzer, J. R. D., M. ¯A. Sfredo, G. D. B. Sousa, J. R. Limaverde. 2007. "Dispersion Coefficient of Coffee Berries in Vibrated Bed Dryer." *Journal of Food Engineering* 79: 905–912. doi: 10.1016/j.jfoodeng.2006.03.011

Gabites, J. R., J. Abrahamson, J. A. Winchester. 2010. "Air Flow Patterns in an Industrial Milk Powder Spray Dryer". *Chemical Engineering Research and Design* 88: 899–910.
Handayani, S. U., M. T. S. Utomo, M. E. Yulianto. 2017. "Performance Evaluation of Continuous Vibrating Fluidized Bed Dryer on Green Tea Production." *Advanced Science Letters* 23: 2530–2532. doi: 10.1166/asl.2017.8676
He, J., Y. Zhao, J. Zhao, Z. Luo, C. Duan, Y. He. 2015. "Separation Performance of Fine Low-Rank Coal by Vibrated Gas–Solid Fluidized Bed for Dry Coal Beneficiation." *Particuology* 23: 100–108. doi: 10.1016/j.partic.2015.02.006
Hsiau, S.-S., C.-H. Tai, M.-C. Chiang. 2004. "Effect of Moisture Content on the Convection Motion of Powders in a Vibrated Bed". *Advanced Powder Technology* 15: 673–686.
Jaraiz, E., S. Kimura, O. Levenspiel. 1992. "Vibrating Beds of Fine Particles: Estimation of Interparticle Forces from Expansion and Pressure Drop Experiments." *Powder Technology* 72: 23–30. doi: 10.1016/s0032-5910(92)85017-p
Jia, D., O. Cathary, J. Peng, X. Bi, C. J. Lim, S. Sokhansanj, Y. Liu, R. Wang, A. Tsutsumi. 2015. "Fluidization and Drying of Biomass Particles in a Vibrating Fluidized Bed with Pulsed Gas Flow." *Fuel Processing Technology* 138: 471–482. doi: 10.1016/j.fuproc.2015.06.023
Lehmann, S. E., E.-U. Hartge, A. Jongsma, I.-M. deLeeuw, F. Innings, S. Heinrich. 2019. "Fluidization Characteristics of Cohesive Powders in Vibrated Fluidized Bed Drying at Low Vibration Frequencies." *Powder Technology* 357: 54–63. doi: 10.1016/j.powtec.2019.08.105
Lehmann, S. E., T. Oesau, A. Jongsma, F. Innings, S. Heinrich. 2020. "Material Specific Drying Kinetics in Fluidized Bed Drying under Mechanical Vibration Using the Reaction Engineering Approach." *Advanced Powder Technology* 31: 4699–4713. doi: 10.1016/j.apt.2020.11.006
Liao, C.-C., and S.-S. Hsiau. 2016. "Transport Properties and Segregation Phenomena in Vibrating Granular Beds." *KONA Powder and Particle Journal* 33: 109–126. doi: 10.14356/kona.2016020
Lima, R. A. B., and M. C. Ferreira. 2011. "Fluidized and Vibrofluidized Shallow Beds of Fresh Leaves." *Particuology* 9: 139–147.doi: 10.1016/j.partic.2010.07.024
Meili, L., R. V. Daleffe, J. T. Freire. 2012a. "Fluid Dynamics of Fluidized and Vibrofluidized Beds Operating with Geldart C Particles." *Chemical Engineering and Technology* 35: 1649–1656. doi: 10.1002/ceat.201100546
Meili, L. R., F. B. Freire, M. C. Ferreira, J. T. Freire. 2012b. "Fluid Dynamics of Vibrofluidized Beds during the Transient Period of Water Evaporation and Drying of Solutions." *Chemical Engineering and Technology* 35: 1803–1809. doi: 10.1002/ceat.201200147
Meili, L., H. Perazzini, M. C. Ferreira, J. T. Freire. 2020. "Analyzing the Universality of the Dimensionless Vibrating Number Based on the Effective Moisture Diffusivity and Its Impact on Specific Energy Consumption." *Heat and Mass Transfer* 56: 1659–1672. doi: 10.1007/s00231-019-02787-8
Mohseni, M., A. Kolomijtschuk, B. Peters, M. Demoulling. 2019. "Biomass Drying in a Vibrating Fluidized Bed Dryer with a Lagrangian-Eulerian Approach." *International Journal of Thermal Sciences* 138: 219 – 234. doi: 10.1016/j.ijthermalsci.2018.12.038
Mujumdar, A. S. 2015. "Principles, Classification, and Selection of Dryers." in *Handbook of Industrial Drying*, 4th ed., edited by A. S. Mujumdar, 3–29. Boca Raton: CRC Press.
Mujumdar, A. S., M. Hasan 2015. "Drying of Polymers." in *Handbook of Industrial Drying*, 4th ed., edited by A. S. Mujumdar, 937–959. Boca Raton: CRC Press.
Mujumdar, A. S., S. V. Jangam, J. Pikón. 2015. "Drying of Coal." in *Handbook of Industrial Drying*, 4th ed., edited by A. S. Mujumdar, 999–1022. Boca Raton: CRC Press.
Nagata, G. A., B. A. Souto, M. T. B. Perazzini, H. Perazzini. 2020. "Analysis of the Isothermal Condition in Drying of Acai Berry Residues for Biomass Application." *Biomass and Bioenergy* 133: 105453e. 10.1016/j.biombioe.2019.105453

Pakowski, Z., A. S. Mujumdar, C. Strumillo. 1984. "Theory and Applications of Vibrated Beds and Vibrated Fluid Beds for Drying Process." in *Advances in Drying*, v.3, edited by A. S. Mujumdar, 245–306. New York: Hemisphere.

Perazzini, H., F. B. Freire, J. T. Freire. 2017. "The Influence of Vibrational Acceleration on Drying Kinetics in Vibro-fluidized Bed." *Chemical Engineering and Processing: Process Intensification* 118: 124–130. doi: 10.1016/j.cep.2017.04.009

Perazzini, H., M. T. B. Perazzini, L. Meili, F. B. Freire, J. T. Freire. 2020. "Artificial Neural Networks to Model Kinetics and Energy Efficiency in Fixed, Fluidized and Vibro-Fluidized Bed Dryers Towards Process Optimization." *Chemical Engineering and Processing: Process Intensification* 156: 108089. doi: 10.1016/j.cep.2020.108089

Picado, A., and J. Martínez. 2012. "Mathematical Modeling of a Continuous Vibrating Fluidized Bed Dryer for Grain." *Drying Technology* 30: 1469–1481. doi: 10.1080/07373937.2012.690123

Raganati, F., R. Chirone, P. Ammendola. 2018. "Gas-solid Fluidization of Cohesive Powders." *Chemical Engineering Research and Design* 133: 347–387. doi: 10.1016/j.cherd.2018.03.034

Rátkai, G. Y., and R. Toros. 1986. "Hydrodynamic Model of the Vibro-Spouted Bed-I. Velocity Profile in the Bed". *Chemical Engineering and Science* 41: 1345–1349. doi: 10.1016/0009-2509(86)87107-x

Rezende, D. R., J. R. D. Finzer. 2008. "Dimensionless on Coffee Fruit Drying in the Vibrated Tray Dryer." *FAZU em Revista* 5: 73–78 (in Portuguese).

Rocha, T. A. F., M. C. Ferreira, J. T. Freire. 2021. "Processing Spent Coffee Ground Powders for Renewable Energy Generation: Mechanical Dewatering and Thermal Drying." *Process Safety and Environmental Protection* 146: 300–311. doi: 10.1016/j.psep.2020.09.003

Strumillo, C., P. L. Jones, R. Zylla. 2015. "Energy Aspects in Drying." in *Handbook of Industrial Drying*, 4th ed., edited by A. S. Mujumdar, 1077–1100. Boca Raton: CRC Press.

Thanimkarn, S., E. Cheevitsopon, J. S. Jongyinhgcharoen. 2019. "Effects of Vibration, Vacuum, and Material Thickness on Infrared Drying of Cissus quadrangularis Linn." *Heliyon* 5: e01999. doi: 10.1016/j.heliyon.2019.e01999

Trávníček, P., J. Novotná, L. Kotek. 2020. Industrial Accidents in Spray Dryer Plants for Dairy Products in Europe. *Journal of Loss Prevention in the Process Industries* 68: 104327.

Yang, X., Y. Zhao, Z. Luo, S. Song, C. Duan, L. Dong. 2013. "Fine Coal Dry Cleaning Using a Vibrated Gas-fluidized Bed". *Fuel Processing Technology* 106: 338–343.

Yazdanpanah, N., and T. A. G. Langrish. 2011. "Crystallization and Drying of Milk Powder in a Multiple-Stage Fluidized Bed Dryer." *Drying Technology* 29: 1046–1057. doi: 10.1080/07373937.2011.561461

Yazdanpanah, N., and T. A. G. Langrish. 2012. "Releasing Fat in Whole Milk Powder during Fluidized Bed Drying." *Drying Technology* 30: 1081–1087. doi: 10.1080/07373937.2012.669000

Yu, X., Z. Luo, H. Li, D. Gan. 2018. The Diffusion Process of the Dense Medium and Its Effects on the Beneficiation of 0–6 mm Oil Shale in the Continuous Vibrating Air Dense Medium Fluidized Bed. *Fuel* 234: 371–383.

Zhao, P., Y. Zhao, Z. Chen, Z. Luo. 2015. "Dry Cleaning of Fine Lignite in a Vibrated Gas-Fluidized Bed: Segregation Characteristics." *Fuel* 142: 274–282. doi: 10.1016/j.fuel.2014.11.029

Zhao, P., L. Zhong, R. Zhu, Y. Zhao, Z. Luo, X. Yang. 2016. "Drying Characteristics and Kinetics of Shengli Lignite Using Different Drying Methods." *Energy Conversion and Management* 120: 330–337. doi: 10.1016/j.enconman.2016.04.105

Zhou, E., Y. Zhang, Y. Zhao, Z. Luo, C. Duan, X. Yang, L. Dong, B. Zhang. 2017. "Collaborative Optimization of Vibration and Gas Flow on Fluidization Quality and

Fine Coal Segregation in a Vibrated Dense Medium Fluidized Bed." *Powder Technology* 322: 497–509. doi: 10.1016/j.powtec.2017.09.034
Zhou, E., Y. Zhang, Y. Zhao, Q. Tian, Z. Chen, G. Iv, X. Yang, L. Dong, C. Duan. 2021. "Influence of Bubbles on the Segregated Stability of Fine Coal in a Vibrated Dense Medium Gas–Solid Fluidized Bed." *Particuology* 58: 259–267. doi: 10.1016/j.partic.2021.03.018

3 Fixed-Bed Drying

Chung Lim Law
Department of Chemical and Environmental Engineering, University of Nottingham Malaysia, Semenyih, Selangor, Malaysia

Hong-Wei Xiao
College of Engineering, China Agricultural University, Beijing, China

Yang Tao
College of Food Science and Technology, Nanjing Agricultural University, Nanjing, Jiangsu, China

CONTENTS

DOI: 10.1201/9781003207108-3

3.1 INTRODUCTION

Fixed-bed drying is a common drying technique. The fundamental theory of a fixed-bed, also known as a packed bed, is well established and we usually learn this topic in particle technology or particle mechanics in chemical engineering or process engineering. Fixed bed is a simple-unit operation and its design is also relatively simple. A typical fixed-bed drying system consists of a fixed-bed dryer chamber, an inlet blower/fan, and a heater/combustor.

As fixed-bed drying is well understood, the recent advancement in fixed-bed drying is more on the design of fixed-bed dryer variants. There are numerous developments on this aspect and it is covered in this chapter.

3.2 FIXED BED/PACKED BED

3.2.1 Packed-Bed Fundamentals

A fixed bed is formed by placing particulate solids in a column, which forms a bed of particles. Figure 3.1 shows a schematic diagram of a fixed bed. It is also known as a packed bed. In fixed-bed drying, the drying medium which is typically in the form of hot air is charged into the fixed-bed dryer from the bottom. When the hot air passes through the packed bed, it makes contact with the particulate solids. Heat and mass transfer occurs. The hot air stream will experience pressure loss when it flows through the perforated distributor placed at the bottom of the fixed bed as well as the bed of particulate solids. A pressure drop across the bed of particulate solids,

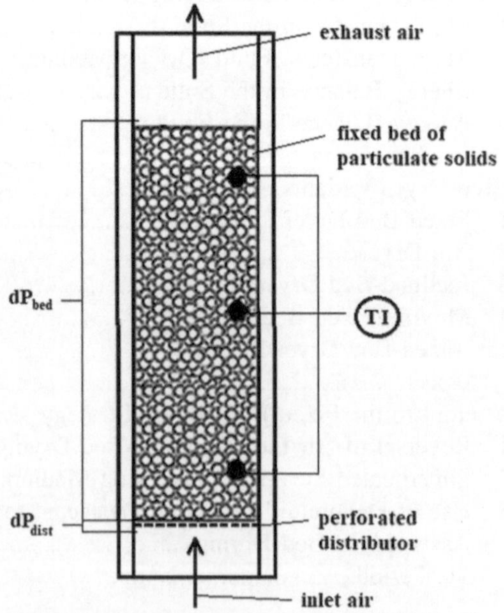

FIGURE 3.1 Fixed-bed operation.

dP_{bed}, and a pressure drop across the distributor plate, dP_{dist}, are important design parameters for fixed-bed dryers.

Heat is transferred from the hot air to the drying material (in this case, the particulate solids). At the initial stage, the heat transfer contributes to the increase of material temperature until it teaches the equilibrium state. At the same time, it also facilitates the evaporation of moisture. When the material temperature reaches the equilibrium temperature, most of the heat transferred from the hot air is used to facilitate drying, which is the evaporation of moisture.

Whilst, mass transfer facilitates the removal of water vapor from the material surface as well as diffusion of internal moisture from the internal to the surface. This is known as liquid diffusion. In certain drying processes, internal moisture may also evaporate internally and the water vapor diffuses internally toward the surface of the material; this is known as vapor diffusion. Drying of porous media or materials with pores and contain internal moisture would need to deal with the diffusion of internal moisture. This is also the main reason why the drying rate towards the end of a drying process is very slow. Diffusion of internal moisture is not easily influenced by external factors such as increasing drying temperature or increasing drying air velocity.

Hot air is charged into the fixed-bed dryer from the bottom; hence the bottom layer of solids undergoes heat and mass transfer in the first instance. If the fixed-bed is thick such as one meter or so, the drying of the fixed-bed may not be uniform. By referring to Figure 3.1, we can install temperature sensors and indicators (TI) at different bed heights in order to monitor the temperature gradient of the entire fixed bed of the particulate solids.

Gaucia et al. (2015) carried out a study on fixed-bed drying of seeds and found that there was a significant drop in the seed drying rate at different bed heights, indicating heterogeneity of the drying potential in different axial positions. Kumoro et al. (2019) reported that in bench-scale fixed-bed drying of paddy, 8 cm of bed height, operated for 60 mins with an air velocity of 5.5 m/s and relative humidity of 37.5%, was the most economical in achieving the final moisture content of 12.8% wb. If this is scaled up, there will be an optimum bed height for fixed-bed drying. Therefore, bed height is an important parameter to be considered if we want to ensure homogeneity in terms of bed properties. Nagle et al. (2008) carried out longan drying using a box-type fixed-bed dryer and reported that both temperature and air velocity profiles were not uniform across the fixed bed due to the design of the air inlet and the fixed-bed dryer chamber. It was reported that fixed-bed drying affected the drying kinetics and product quality. They also observed that high humidity and convective cooling of the drying air passing through the fixed bed of longan has resulted in condensation on longan surfaces in the upper layer at the initial stage of drying.

If the pressure drop across the bed of particulate solids is measured when the hot air velocity is increased from zero, the bed pressure drop will rise and eventually reach a maximum value. Figure 3.2 shows a typical bed pressure drop versus hot air velocity.

With reference to Figure 3.2, the region between 0-A is the flow regime of a packed bed or a fixed bed. Depending on the materials that form the fixed bed, 0-A can be a linear function or a quadratic function indicated by 0-A'-A.

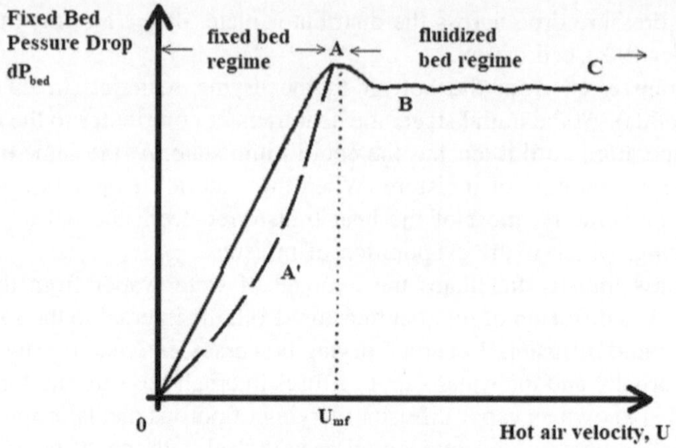

FIGURE 3.2 Pressure drop across a bed of particulate solids.

The relationship between the bed pressure drop and hot air velocity can be described by the Ergun equation, which is shown in Eq 3.1:

$$\frac{(-\Delta P)}{H} = 150\frac{\mu U(1-\epsilon)^2}{d_p^2\epsilon^3} + 1.75\frac{\rho_f U^2}{d_p^2}\frac{(1-\epsilon)}{\epsilon^2} \ldots \tag{3.1}$$

At low hot air velocity, such as 0.1 m/s, where the flow is laminar, the first term of the Ergun equation is dominant and hence the second term is negligible. Hence, it becomes a linear function where the relationship between the pressure drop and the hot air velocity is linear, and gives us a straight line in a (–ΔP) vs U plot. When the hot air velocity becomes higher, such as higher than 1 m/s, the second term becomes dominant. Then the Ergun equation becomes a quadratic function and gives us a quadratic curve in a (–ΔP) vs U plot. When the hot air velocity is increased further, the pressure drop remains constant.

3.2.2 Heat and Mass in Packed-Bed Drying

When a fixed bed of drying material is exposed to hot air, heat is transferred from the hot air to the drying material. Meanwhile, surface moisture on the drying material is vaporized and internal moisture is migrated to the material surface. Therefore fixed-bed drying involves both heat transfer and mass transfer.

3.2.2.1 Mass Transfer in the Solid

Moisture content of the drying material is decreased over time, and known as the drying rate.

$$f = \frac{\partial X}{\partial t} \ldots \tag{3.2}$$

where X is material moisture content in dry basis $\left(\frac{kg_{H2O}}{KgDS}\right)$ and f is the drying rate $\left(\frac{kg_{H2O}}{s\,.\,kgDS}\right.$ or simplified to $\left.\frac{1}{s}\right)$. DS is a dry solid.

3.2.2.2 Mass Transfer in Fluid (Drying Medium)

Moisture loss during the drying process is picked or absorbed by drying air.

$$\rho_b\left(-\frac{\partial X}{\partial t}\right) = \rho_g\left(\frac{v_g}{S_b}.\frac{\partial Y_g}{\partial \gamma} + \frac{\partial Y_g}{\partial t}\right)\ldots \tag{3.3}$$

where ρ_b is solid bulk density $\left(\frac{kgDS}{m^3}\right)$, ρ_g is fluid density $\left(\frac{kgDA}{m^3}\right)$, DA is dry air, v_g is air velocity $\left(\frac{m}{s}\right)$, S_b is shrinkage parameter (m), Y_g is air humidity $\left(\frac{kgH2O}{kgDA}\right)$, and γ is dimension-less moving coordinate (–).

3.2.2.3 Energy Balance in the Solid

Heat transfer from the drying medium (hot air) to the drying material is absorbed by the solid, which results in a solid temperature increase and increase of moisture temperature (sensible heat) as well as vaporization (latent heat).

$$ha_v\left(T_g - T_S\right) = \rho_b\left(c_{ps} - c_{pw}\right)\frac{\partial T_S}{\partial t} + \rho_L\left[L_p + \left(c_{pv} - c_{pw}\right)T_S\right]\left(-\frac{\partial X}{\partial t}\right)\ldots \tag{3.4}$$

where h is heat transfer coefficient $\left(\frac{J}{m^2 s\,^\circ C}\right)$, a_v is specific surface area $\left(\frac{1}{m}\right)$, T is temperature (°C), c_p is specific heat $\left(\frac{J}{kg\,^\circ C}\right)$, and L_p is latent heat $\left(\frac{J}{kg}\right)$. Subscript v denotes vapor, w is water, and S is solid.

3.2.2.4 Energy Balance in the Fluid

Heat transfer from the drying medium (hot air) to the drying material is absorbed by the solid, which results in the decrease of hot air temperature, including the hot air and vapor within the packed bed, and the increase of moisture temperature in the solid (sensible heat).

$$ha_v\left(T_g - T_S\right) = \frac{v_g\rho_g}{S_b}\left(c_{pg} + Y_g c_{pv}\right)\left(-\frac{\partial T_g}{\partial \gamma}\right) + \rho_g\varepsilon\left(c_{pg} - c_{pv}\right)\left(-\frac{\partial T_g}{\partial t}\right)$$

$$+ \rho_b c_{pv}\left(-\frac{\partial X}{\partial t}\right)\left(T_g - T_S\right)\ldots \tag{3.5}$$

ε is bed porosity (–). Subscript g denotes hot air and v is vapor.

Prado and Sartori (2008, 2011) studied the heat and mass of packed-bed drying of porous material and took into account the effect of material shrinkage in the heat and mass transfer model. They reported that bed porosity (ε) increases and solid specific surface area (a_v) decreases as the packed drying is carried out. Solid temperature increases and stays constant when it reaches the equilibrium; the temperature profile is a sigmoid curve. Across the bed height, there is a significant temperature gradient where the solid at the bottom of the packed bed has a higher temperature compared to the solid at the top layer of the packed bed.

3.3 FIXED-BED DRYER AND ITS VARIANTS

Drying can be conducted in any of the flow regimes (refer to Figure 3.2) and a fixed bed is carried out at a lower hot air velocity compared to other flow regimes (such as fluidized bed). Fixed-bed drying has the following advantages:

- Lower air velocity; hence, lower operating cost in term of operating the blower that supplies air into the fixed bed
- As the bed of particulate solids remains in the bed, there is no issue with the entrainment of fine particulate solids with the entraining air stream; hence, there is no need to install a solid air separation system at the exhaust
- Since there is no particulate solids movement, hence, there is no particle attrition due to collision
- The whole fixed-bed drying system is simple and easier to maintain, and the maintenance cost is relatively lower

At the same time, a fixed bed also gives limitations:

- As particulate solids are not moving within the bed, certain sections of the bed may not be exposed to the hot air stream; hence, poor heat and mass transfer. This in turn results in uneven drying performance across the entire bed of particulate solids.
- Likewise, certain sections may be exposed to high degrees of heat and mass transfer; hence, the drying of these sections is more intense and may result in over-drying and overheating. Hot spots may occur as a result. Therefore, it causes fixed-bed drying to suffer from large undesirable thermal gradient across the fixed bed. Due to the formation of hot spots and other sections that are not exposed to hot air, temperature control of fixed-bed drying is relatively poor.
- If the fixed bed consists of fine powder, the fixed bed may give channeling and hot air stream may only flow through the channels instead of making contact with every particulate solid in the bed. This results in poor heat and mass transfer and the hot air stream is seen to have bypassed the fixed bed. Hence, the drying performance is extremely poor if bypassing of drying air occurs.

Gazor and Alizadeh (2020) carried out rotary drying and fixed-bed drying of a paddy and reported that fixed-bed drying of a paddy exhibited non-uniformity in terms of temperature profile across the fixed bed. They reported that the lower layer of paddy experienced higher thermal stress, which affected the head rice yield when the paddy was subjected to milling.

3.3.1 Fixed-Bed Dryer Variants

The basic principles of fixed-bed operation are simple and hence a fixed-bed drying system is not complex. However, due to its inherent limitations mentioned above, there are many different designs of fixed-bed dryers that have been accomplished by many design engineers and researchers. The intention was to overcome the limitations of a fixed-bed operation.

According to Law and Mujumdar (2008), the type of dryer can be classified based on four main factors. They are:

- Drying strategy – the strategy of how we accomplish drying. We may consider i. performing drying at a low temperature e.g. freeze drying, vacuum drying, or heat pump assisted drying; or ii. performing drying at an intermittent mode or varying the operating parameter, such as intermittent drying, pulse combustion drying, or cyclic pressure drying; or iii. exposing the drying material with a drying medium (hot air) using an unconventional method such as impinging stream drying
- Drying medium – the type of drying medium that we can use in a drying operation. We may consider i. using hot air e.g., hot air drying, convective drying; ii. harnessing solar energy to produce hot air such as sun drying or solar drying; iii. using superheated steam e.g. superheated steam drying; iv. using a heat pump system to generate a low temperature (in the range of −10°C to 40°C) and low relative humidity drying medium e.g., heat pump assisted drying; v. using insert solids to assist in drying; vi. using desiccant to generate a low relative humidity drying medium; vi. carrying out drying using supercritical fluid technology, e.g. supercritical fluid drying
- Ways of handling drying material in a drying chamber; for instance, drying materials stay stationary, such as tray dryer, fixed-bed dryer; stirring, agitating, or rotating the drying materials such as drum dryer, agitating dryer, tumbler dryer, or paddle dryer; fluidize the drying materials, such as fluidized bed dryer, spouted bed dryer, vibrating bed dryer, pulsating bed dryer, or jet zone dryer
- Mode of heat input, namely convection, such as hot air dryer, superheated steam dryer, or modified atmosphere dryer; conduction such as pan dryer or paddle dryer; and radiation such as infrared dryer or dielectric dryer including radio frequency and microwave dryers.

Figure 3.3 shows the classification modified from law and Mujumdar (2008).

3.3.1.1 Fixed-Bed Dryer

As described above, a fixed-bed dryer is simple. Figure 3.4 shows a typical fixed-bed drying system. It consists of a packed-bed column, an inlet blower, and a heater/combustor and may be an outlet air blower/fan if there is a need to install a gas-solid separation system (such as cyclone or bag filter) at the outlet.

Atmospheric air is sucked by a blower at the inlet, goes through a heater, and its temperature is elevated. The hot air is then charged into the fixed-bed dryer chamber and makes contact with the fixed bed of particulate solids. After the heat and mass transfer take place in the drying chamber, the exhaust air is sucked by an outlet blower. A solid gas separation system can be installed in order to remove suspended solids in the exhaust air before it is discharged into the atmosphere.

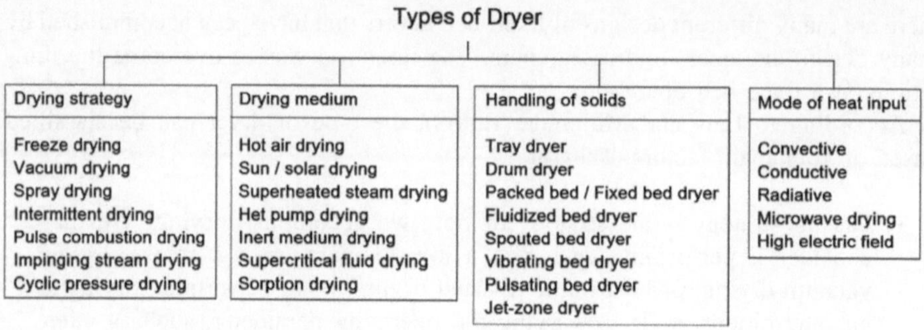

FIGURE 3.3 Classification of dryers.

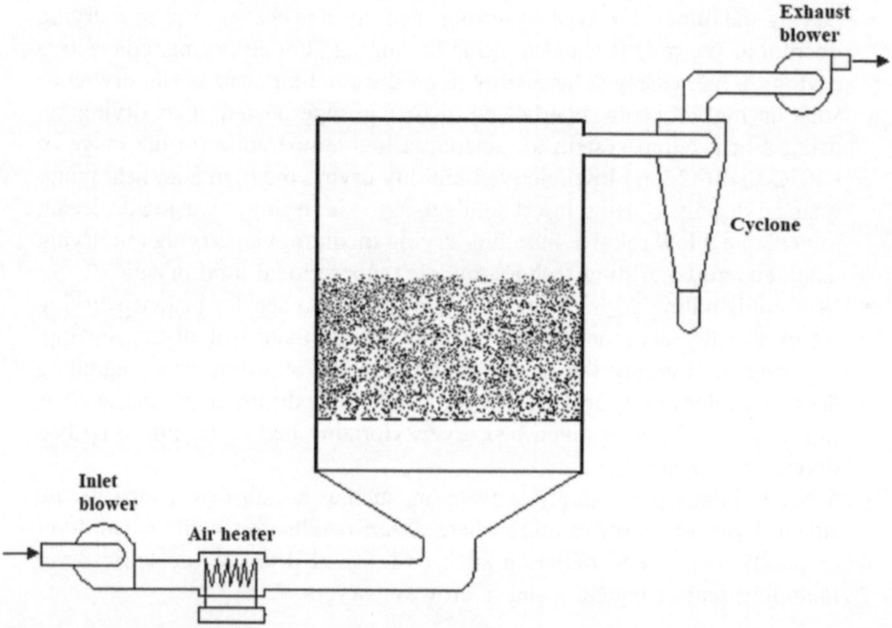

FIGURE 3.4 A typical fixed-bed dying system.

Instead of using a heater that depends on electricity, we can harness solar energy to elevate the temperature of the inlet air stream. Figure 3.5 shows a solar dryer where the drying chamber is designed to be a multiple fixed bed. A solar air collector can be installed at the inlet and an inlet fan/blower pushes the heated air into the drying chamber. If the heated air generated by the solar collector does not meet the design temperature, an electric-powered heater can be installed to further increase the temperature to the desired value.

If a deep fixed bed poses some difficulty in ensuring good contacting efficiency between the drying medium and the drying material, the fixed bed can be segregated or divided into a number of swallow beds (Figure 3.5). This is known as multiple

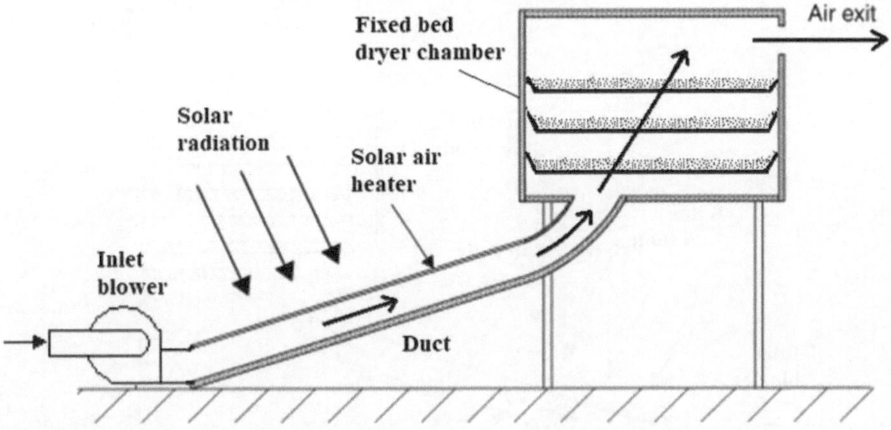

FIGURE 3.5 Solar fixed-bed dryer, also known as solar tray dryer.

fixed beds. Figure 3.6 shows another example of a multiple fixed-bed dryer, where it uses desiccant and solar energy to generate the drying medium. Padmanaban et al. (2017) applied desiccant-assisted fixed-bed solar dryer to dry copra. They reported that the total drying was 50% lesser compared to a conventional solar dryer. This is due to the fact that the packed-bed absorber plate configuration was able to retain the heat for a few more hours, which allowed the drying to be prolonged, even though solar radiation was not available. In addition, the desiccant is able to reduce the relative humidity of the drying air; therefore, the drying air has a higher capacity to absorb more moisture from the drying material.

Figure 3.7 shows another design of a multiple fixed bed where the air flow is designed to flow within the drying chamber in a zig-zag manner. This would allow

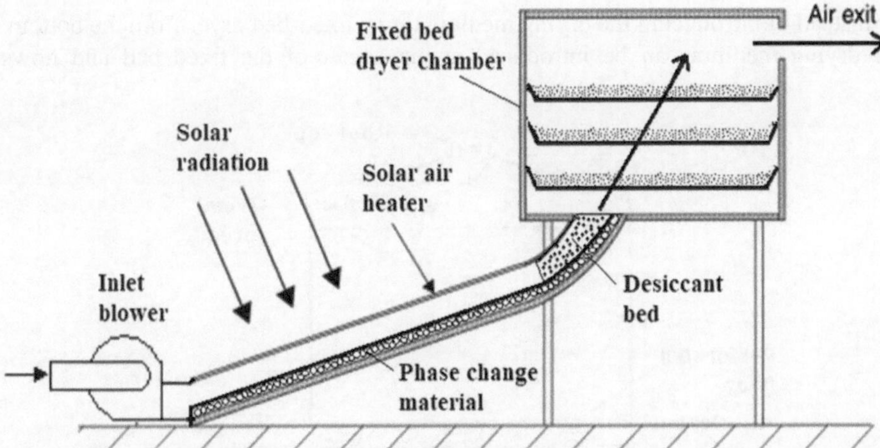

FIGURE 3.6 Multiple fixed-bed using desiccant and solar energy as a heating source.

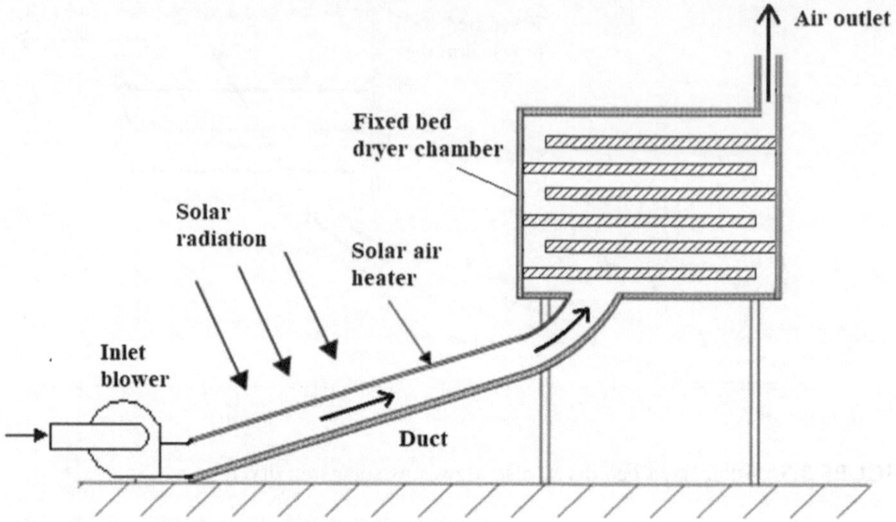

FIGURE 3.7 Solar assisted multi-stage fixed-bed dryer.

the drying medium to stay longer in the drying chamber and have more opportunity to make contact with more particulate solids in the multi-stage fixed bed of particulate solids.

3.3.1.2 Bin Dryer

A bin dryer is especially popular in the grain industry. The bin dryer is huge in size and therefore the grains form a deep bed. An optimum height of the bed of grain is about 1 m ± 30 cm. The bed height can be increased to 2–3 m with the installation of a stirring device. Figure 3.8 shows a typical bin dryer that is used in the grain industry.

Instead of introducing the drying medium into a fixed-bed dryer from the bottom, the drying medium can be introduced at the center of the fixed bed and flows

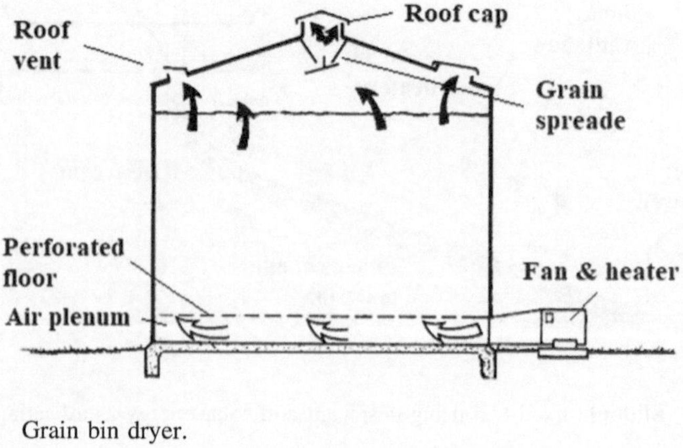

FIGURE 3.8 Grain bin dryer.

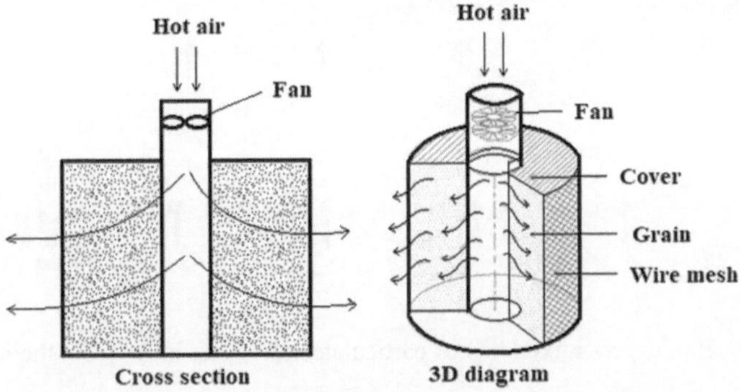

FIGURE 3.9 Circular bin dryer.

outwardly in a radial direction. Figure 3.9 shows a typical design of a circular bin dryer. It is more compact compared with a bin dryer and saves floor space. However, its contacting efficiency is less than a bin dryer.

3.3.1.3 Inclined-Bed Dryer

If the fixed-bed dryer capacity is large, unloading the dried materials from the fixed-bed dryer might be troublesome and laborious. This is because the fixed bed is flat and to empty the fixed-bed dryer chamber completely, workers have to be deployed to remove the materials after the fixed-bed drying process. If the fixed bed is tilted at one end, it becomes an inclined bed. An inclined bed can facilitate the unloading of dried material from the dryer using the gravitational force. An inclined bed is used in a paddy/rough rice milling facility.

Sarker et al. (2014) carried out an inclined-bed drying paddy and reported that the grains located at the air inlet (bottom layer) dried faster than the paddy at the air outlet (top layer). This resulted in a moisture gradient across the bed of the paddy in the inclined bed. At the beginning of the drying, the moisture removed from the bottom travels through the bed of the paddy and the middle layer as well as the top layer may absorb some moisture. At the end of drying, when the top layer of the paddy is dried, the bottom layer might be over-dried. If this occurs, the paddy at the bottom may experience physical stress due to over-drying and results in cracking of the rice kernels. Due to its sensitivity to thermal stress, inclined-bed drying of a paddy is typically carried out at about 41°C–42°C.

Ghiasi et al. (2016) compared the performance of a flat-bed dryer and inclined-bed dryer in the drying of a paddy. They reported that the inclined-bed dryer significantly increased the drying capacity (ton/m^{2}h) of up to 25% at drying temperatures of 42°C–43°C and almost 29% at drying temperatures of 38°C–39°C compared to flat-bed drying. Further, an inclined bed recorded an overall drying energy consumption of 78.6–91.97 kWh/ton, which is much lower than that of a flat-bed dryer, which recorded 200 kWh/ton. On the other hand, a fixed-bed dryer gave slightly better head rice yield: 58% versus 52%.

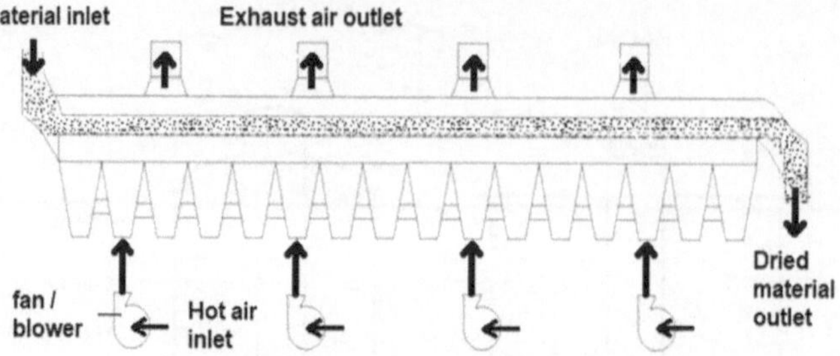

FIGURE 3.10 Belt dryer, a fixed bed of particulate solids that move from the dryer inlet to the outlet.

3.3.1.4 Moving-Fixed Bed Dryer

Contacting efficiency between drying medium (e.g., hot air) and drying materials in a fixed bed may be poor, especially if the drying material is in the form of fine powder. In this regard, the fixed bed can be moved and transported from one point to another while the particulate solids remain stationary. Hence, the bed of particulate solids is still a fixed bed (normally a swallow bed) while it is exposed to the drying medium.

Figure 3.10 shows an example of a moving-bed dryer. Particulate solids are normally placed on a conveyor belt or wire mesh while the particulate solids are transported from the inlet point to the discharge point.

3.4 ADVANCEMENT IN FIXED-BED DRYER RESEARCH

3.4.1 Quality Aspect

Product quality is the main aspect that we need to consider when it comes to dryer selection and determining the operating parameter. Rapid drying may be desirable if it impacts the quality of the dehydrated product.

Glaucia et al. (2015) carried out fixed-bed drying of soybeans and reported that the best germination and vigor indices were attained at low values of air temperature (lower than 35°C) coupled with a high air relative humidity (higher than 35%). In addition, high air relative humidity (higher than 35%) and low seed temperature assure better seed physical quality, expressed by the high index of non-fissured seeds. This indicates rapid drying may not be desirable as it has an adverse impact on the product quality. In general, the decrease in seed quality is associated with conditions that lead to a high drying rate and corresponding over-drying conditions.

Leyva Daniel et al. (2012) performed fixed-bed drying of lixiviated roselle calyxes and found that a fixed-bed dryer is a suitable drying technique for drying the material, as the results showed that the retention of monomeric anthocyanins and phenolic compounds were satisfactory and of course, provided that the fixed-bed drying was performed at a relatively lower drying temperature.

Thakur and Gupta (2006) carried out fixed-bed drying and fluidized bed drying of a paddy. They reported that both fixed-bed and fluidized-bed drying gave similar results in terms of head rice yield. However, Luthra and Sadaka (2021) reported that a fixed-bed dryer gave better head rice yield than a fluidized-bed dryer when they used silica gel as a dehumidification agent to assist both fixed-bed and fluidized-bed drying. Fluidized-bed drying gave a faster drying kinetics; however, it is advisable to avoid rapid drying when using a fluidized bed at a higher fluidizing gas velocity. According to Thakur and Gupta (2006), the fluidized-bed drying should be carried out at a minimum fluidization regime, where it is close to the fixed-bed maximum operating velocity. It was reported that energy savings of 20%–50% (depending on the tempering duration) could be achieved by introducing tempering intermittently. In addition, tempering could also increase the head rice yield.

Braga et al. (2005) carried out fixed-bed drying of long piper leaves and reported that the essential oil extracted from the piper leaves is heat sensitive; therefore, the drying temperature cannot exceed 50°C. They concluded that 45°C is optimum in performing fixed-bed drying of long piper leaves in order to maximize the retention of essential oil in the dehydrated piper leaves. In another fixed-bed drying experimental work, Braga et al. (2020) reported that sacaca and alfavaca leaves also gave similar findings and, further, sacaca required a much lower air velocity and drying temperature.

3.4.2 New Attempt to the Fixed-Bed Drying Strategy

3.4.2.1 Reversal of Air Flow in Fixed-Bed Drying

Albini et al (2018) attempted the reversal of air flow periodically and reported that it could reduce the moisture gradients across the fixed bed of barley and gave a better bed homogeneity. Anyway, reversing the air flow caused a slight decrease in drying rates because moisture removed from the top layer of the fixed bed (which was wetter) was captured by the bottom layer, which was drier when the drying air flows from the top to bottom in the reverse mode. Ratti and Mujumdar (1995) simulated the reverse mode in packed-bed drying of carrots and found that it could flatten the temperature gradient and moisture gradient across the packed bed. This enabled uniformity in product quality. This is especially important to a thick packed bed as the temperature and moisture would be very large. Herman-Lara et al. (2010) did the same work on carrots and reported that it could increase the drying rate and reduce the energy consumption. This is applicable to a bed height that is thicker and the packed-bed drying takes a longer time, especially if the falling rate period is long. Even though the bottom layer may absorb some moisture when it is operated in reverse mode, the moisture could be easily removed when the air flow is switched back to the normal mode.

3.4.2.2 Superheated Steam as the Drying Medium

Hot air can be replaced by another drying medium, such as nitrogen, superheated steam, etc. Tran (2020) reported the work on superheated steam-packed bed drying of ceramic particles and concluded that a superheated steam is a suitable drying medium for fixed-bed drying after looking at the simulation of moisture profile and particle and vapor temperature profiles during the fixed-bed drying.

3.4.2.3 Use of Dehumidification and Desiccant to Assist Fixed-Bed Drying

A drying medium at the dryer inlet may be dehumidified in order to lower its relative humidity. This is to increase the drying capacity of the drying medium. Dehumidification is especially useful when we want to conduct the fixed bed at a low temperature. In a typical heating process, the drying medium heated to a relatively lower temperature (i.e., 40°C–50°C) tends to have higher relative humidity and its drying capacity is less compared to a drying medium at a higher temperature. Luthra and Sadaka (2021) carried out fixed-bed and fluidized-bed drying of a paddy using silica gel as a dehumidification agent and they confirmed that using a desiccant is a good way to dehumidify the drying medium and it gave positive results to the head rice yield of the dried paddy.

3.4.3 Drying Rate Period

Sghaier et al. (2009) carried out fixed-bed drying of moist porous alumina particles and showed that when the drying material is moist, a constant rate period exists and the drying material temperature is at the corresponding wet bulb temperature. When the drying material became partially dried and especially when the surface is dry, the drying process entered into a falling rate period. The temperature of the drying material increased until it reached an equilibrium value. If the drying air has a higher relative humidity, the corresponding wet bulb temperature will be higher.

Pusat and Erdem (2017) carried our fixed-bed drying of course lignite particles and reported that the falling rate period prevailed in the entire drying period. The existence of a falling rate period indicates that diffusion governed the whole transport process of internal moisture removal during the fixed-bed drying of coarse lignite particles.

3.5 CONCLUSION

A fixed bed has its own advantages over other types of drying techniques and also has its own limitations. A good understanding of its advantages and its limitations would allow us to make a good decision in selecting the proper drying technique for our unit operation. Even though after we have decided to use a fixed-bed drying for our drying unit operation, we may still need to select the fixed-bed dryer variant that is the most suitable for our drying process.

REFERENCES

Albini, G., Freire, F. B., Freire, J. T. 2018. Barley: Effect of airflow reversal on fixed-bed drying. *Chemical Engineering and Processing – Process Intensification*, 134: 97–104. 10.1016/j.cep.2018.11.001

Braga, N. P., M. A. Cremasco, Valle R. C. C. R. 2005. "The Effects Of Fixed-Bed Drying on the Yield and Composition of Essential Oil from Long Pepper (*Piper hispidinervium* C. DC) Leaves". *Brazilian Journal of Chemical Engineering* , 22: 257–262.

Braga, N., Veiga Jr V. F., Chaves Francisco C. M., Oka Jaisson M., Lima Milena C. F. 2020. Effects of convective drying in fixed-bed on content and composition of essential oils from Croton cajucara Benth and Ocimum micranthum Willd. *International Journal of Development Research*, 10(7): 37776–37781. 10.37118/ijdr.19216.07.2020

Braga, N. P., Cremasco, M. A., Valle, R. C. C. R. 2020. The effects of fixed-bed drying on the yield and composition of essential oil from long pepper (*Piper hispidinervium* C. DC) leaves. *Brazilian Journal of Chemical Engineering*, 22(2): 257–262.

Gazor, H. R., Alizadeh, M. R. 2020. Comparison of rotary dryer with conventional fixed-bed dryer for paddy drying, milling quality and energy consumption. *CIGR Journal*, 22(2): 264-271.

Ghiasi, M., *Ibrahim, M. N., Kadir Basha, R., Abdul Talib, R. 2016. Energy usage and drying capacity of flat-bed and inclined-bed dryers for rough rice drying. *International Food Research Journal*, 23: S23–S29.

Glaucia, F. M. V., Souzaa, Ricardo, F., Mirandab, Marcos A. S. Barrozoa. 2015. Soybean (Glycine max L. Merrill) seed drying in fixed-bed: Process heterogeneity and seed quality. *Drying Technology*, 33(14): 1779–1787. 10.1080/07373937.2015.1039542

Herman-Lara, E., Martinez-Sanchez, C. E., Amador-Mendoza, A. A., Ruiz-Lopez, I. I. 2010. Effect of airflow reversal on packed-bed drying of carrots. *Journal of Food Process Engineering*, 33: 684–700. 10.1111/j.1745-4530.2008.00296.x

Kumoro, A. C., Lukiwati, D. R., Praseptiangga, D., Djaeni, M., Ratnawati, R., Hidayat, J. P., Utari, F. D. 2019. Effect of bed thickness on the drying rate of paddy rice in an up-flowfixed-bed dryer. *Journal of Physics: Conference Series* (Vol. 1376, No. 1, p. 012045). 1376 012045. IOP Publishing.

Law, C. L., Mujumdar, A. S. (2008). Dehydration of fruits and vegetables. In A. S. Mujumdar (Ed.), *Guide to industrial drying* (pp. 223–249). India: Three S Colors.

Leyva Daniel, D., Barragán Huerta, B. E., Anaya Sosa, I., Vizcarra Mendoza, M. G. 2012. Effect of fixed-bed drying on the retention of phenolic compounds, anthocyanins and antioxidant activity of roselle (Hibiscus sabdariffa L.). *Industrial Crops and Products*, 40: 268– 276. 10.1016/j.indcrop.2012.03.015

Luthra, K., Sammy, S. 2021. Investigation of rough rice drying in fixed and fluidized bed dryers utilizing dehumidified air as a drying agent. *Drying Technology*, 39(8): 1059–1073. 10.1080/07373937.2020.1741606

Nagle, M., González-Azcárraga, J. C., Phupaichitkun, S., Mahayothee, B., Haewsungcharern, M., Janjai, S., Leis, H., Müller, J. 2008. Effects of operating practices on performance of a fixed-bed convection dryer and quality of dried longan. *International Journal of Food Science and Technology*, 43: 1979. 10.1111/j.1365-2621.2008.01801.x

Padmanaban, G., Palani, P. K., Murugesan, M. 2017. Performance of a desiccant assisted packed bed passive solar dryer for copra processing. *Thermal Science*, 21(2) S419–S426. 10.2298/TSCI17S2419P

Prado, M. M., Sartori, D. J. M. 2008. Simultaneous heat and mass transfer in packed bed drying of seeds having a mucilage coating. *Brazilian Journal of Chemical Engineering*, 25(1): 39–50.

Prado, M. M., Sartori, D. J. M. 2011. Heat and mass transfer in packed bed drying of shrinking particles. In Mohamed El- Amin (Ed.), *Mass Transfer in multiphase systems and its applications* (pp. 621–648), Croatia, Shanghai: InTech.

Pusat, S., Erdem, H. H. 2017. Drying characteristics of coarse low rank coal particles in a fixed-bed dryer. *International Journal of Coal Preparation and Utilization*, 37(6): 303–313. 10.1080/19392699.2016.1179638

Ratti, C., Mujumdar, A. S. 1995. Simulation of packed bed drying of foodstuffs with airflow reversal. *Journal of Food Engineering*, 26: 259–271. 10.1016/0260-8774(94)00007-V

Sarker, M. S. H., Nordin Ibrahim, M., Ab. Aziz, N. and Mohd. Salleh, P. 2014. Energy and rice quality aspects during drying of freshly harvested paddy with industrial inclined bed dryer. *Energy Conversion and Management*, 77: 389–395. 10.1016/j.enconman.2013.09.038

Sghaier, J., Messai, S., Lecomte, D. and Belghith, A. 2009. High temperature convective drying of a packed bed with humid air at different humidities. *American Journal of Engineering and Applied Sciences*, 2(1): 61–69. 10.3844/ajeassp.2009.61.69

Thakur, A. K., Gupta, A. K. 2006. Stationary versus fluidized-bed drying of high-moisture paddy with rest period. *Drying Technology*, 24(11): 1443–1456. 10.1080/07373930600952792

Tran, T. T. H. 2020. Modelling of drying in packed bed by superheated steam. *Journal of Mechanical Engineering Research and Developments*, 43(1): 135–142.

4 Pneumatic and Flash Drying

Masoud Dorfeshan
Department of Mechanical Engineering, Behbahan Khatam Alanbia University of Technology, Behbahan, Iran

Salem Mehrzad
Department of Mechanical Engineering, Engineering Faculty, Shahid Chamran University of Ahvaz, Ahvaz, Iran

CONTENTS

4.1 INTRODUCTION

It is well known that the drying processes are highly energy intensive. Hence, improving the design of existing dryers can significantly reduce the operating cost of drying systems. A large variety of dryers are developed for various industrial

DOI: 10.1201/9781003207108-4

applications, most with alterations to make them suitable for target industries. The initial moisture content in the feed material and other effects, such as agglomeration and attrition, must be considered while choosing a suitable dryer. Both result in changes to particle size and particle shape that can impact post-drying steps. The properties of the final product in terms of the state of the material and the degree of adhesion determine what type of dryers can be used. Many solid particles that need to be dried are porous in nature (Aubin et al., 2014). When the whole particle drying occurs in a pneumatic dryer, the process can be divided into three stages: preheating (warming-up), constant-rate drying, and falling-rate drying.

In the preheating stage, the particle temperature rises. In constant-rate drying, excess moisture around the particle evaporates. This action is similar to drop evaporation and decreases the diameter of the wet particle, so-called outer diameter contracts. The surface temperature of the solid reaches the wet-bulb temperature corresponding to the air humidity and temperature conditions at a similar location. As soon as the surface moisture evaporates, the falling-rate drying begins. The wet particle has two different regions at this stage: the dry crust layer and the wet core region. The drying rate at this stage is poorly controlled based on the infiltration of moisture from the pores inside the porous core to the particle's outer surface. By transferring moisture from the core to the shell, the particle shrinks, and the thickness of the dry shell increases. In the falling-rate period, the particle's temperature reaches the dry-bulb temperature of drying gas. These issues must be carefully addressed when drying heat-sensitive materials. Typically, mathematical modeling of the second stage is performed by considering the constant particle diameter (Mezhericher, Levy and Borde, 2010). In a pneumatic dryer, gas flow, for example, hot air flow or superheated steam, and the flow of conveying particles, is also responsible for drying. The hot drying medium can be provided by other existing processes, or separately. It can be supplied through a heating system or a superheated steam generation system.

4.2 PRINCIPLE OF PNEUMATIC DRYING PROCESSES

Pneumatic drying is widely used in various industries, and the process is a combination of heat and mass transfer, and pneumatic conveying technology. The existence of surface mass transfer and high heat transfer have led to these dryers having a high drying rate and high drying capacity (Skuratovsky et al., 2005).

Pneumatic drying is a process in which particles can be injected into a stream of hot air using hot air and also by the injection system because the material dries quickly. Due to the small size of the particles, it is possible to dry fast, and particles' residence time is very short, in the range of a few seconds (Tanaka et al., 2008).

Despite the certain simplicity of the process, pneumatic drying is a complex multiphase transfer phenomenon involving compact turbulent flow and multicomponent wet particles. It also includes the heat and mass transfer between the drying gas and the wet particles, the drying kinetics of the particles, and the mechanical and thermal stresses within the particles (Mezhericher et al., 2010).

Pneumatic drying is similar to pneumatic conveying hydrodynamically (Aubin et al., 2014). Based on the mechanism of particle stransfer, drying equipment is divided into batch and continuous categories (El-Behery et al., 2012). Pneumatic dryers are considered continuous types. The design of a pneumatic dryer is based on the properties of gas and particles and hydrodynamics and thermodynamics of the drying process. The thermal efficiency of pneumatic dryers is in the 50%–75% range (Crapiste and Rotstein, 1986). A high-temperature drying fluid should be used to improve efficiency until the product is not overheated and thermally damaged.

The heat transfer potential between the gas and the particles is higher at three zones: the feeding point, the elbows, and the cyclone (Debrand, 1974). This potential can be explained by the higher slip velocity and the high sensitivity of the turbulence phenomenon. There is an oncoming temperature drop along the pneumatic dryer and causes a significant reduction in gas velocity. Also, this velocity decrease is a function of gas density and gas viscosity.

Particle velocity variation can change the drying rate, resulting in heat and mass transfer coefficients, and the interaction of these parameters should be noted. These problems are implemented and used in the design of mathematical models used in the simulation of the pneumatic driving process (Shigeru and Pei, 1984).

4.3 ADVANTAGES AND LIMITATIONS

Pneumatic drying can be quite interesting, technically and economically. Some advantages of this technique are:

- Highest moisture removal rate (Indarto et al., 2007), so they are more thermally efficient
- Simple structure and low construction cost
- Short residence time that allows drying the heat-sensitive particles
- Parallel flow that allows the use of high-temperature drying gas at the inlet leads to high thermal efficiency
- High drying capacity due to continuous process
- Can be used as a conveying system
- Easy installation and easy maintenance due to few moving parts

The short residence time can, however, be a complication and an advantage. The disadvantages of this technique are:

- Not suitable for brittle and highly adhesive materials
- Short residence time limits its scope of use
- Used to separate surface moisture mainly and does not provide time to separate bound moisture

4.4 DESIGN AND OPERATION

An experimental method is usually required to study the performance of a pneumatic drying system. Therefore, a preliminary acquaintance with the mechanical components and effective parameters in the analysis seem necessary. Designing a pneumatic dryer requires a variety of processed data, presented in Table 4.1.

TABLE 4.1
Required Data for Pneumatic Dryer Design (Baeyens et al., 1995)

Data Category	Required Data
Thermal	• Capacity
	• Feed solid temperature
	• Material thermal properties
Hydrodynamic	• Air properties
	• Mean particle diameter
	• Particle envelope density
	• Particle abrasivity
	• Particle hardness and resistance to attrition
Psychrometric	• Air inlet moisture content
	• Solid inlet moisture content and critical moisture content
Geometric	• e.g., maximum length for the space available
Drying thermodynamics	• Inlet and outlet gas temperature
Drying hydrodynamic	• Choking velocity
	• Diameter of the duct
	• Air-flow rate
	• Air velocity
	• Safety factor
	Pressure drop
Drying psychrometric	Heat transfer coefficient
	Required duct length

The schematic diagram of a simple flash dryer is shown in Figure 4.1. In general, the flow path can be indirect and have diameter changes.

In the following section, more information about the main components of this system is provided.

4.4.1 Air Supply System

Drying gas velocity is a very important factor in the pneumatic drying process. If the gas velocity is low, it cannot carry the particles, and the particles may settle and block the parts of the path. On the other hand, if the gas velocity is too high, the particle residence time will be shorter, and there will not be enough time for heat and moisture exchange between the gas and the particles.

Various equipment is used for air supply in pneumatic dryers. Fan and blower are good options if a high flow rate and low pressure are required. In systems that require high pressure for long-distance or dense phase conveying, one of the types of a compressor can be used.

4.4.2 Heating System

Heat input to the pneumatic dryer is typically by combustion of natural gas or indirect steam heat through steam coils. However, several other heat sources are used. Some other common heat sources are LP gas, fuel oil, thermal oil, or electric heaters.

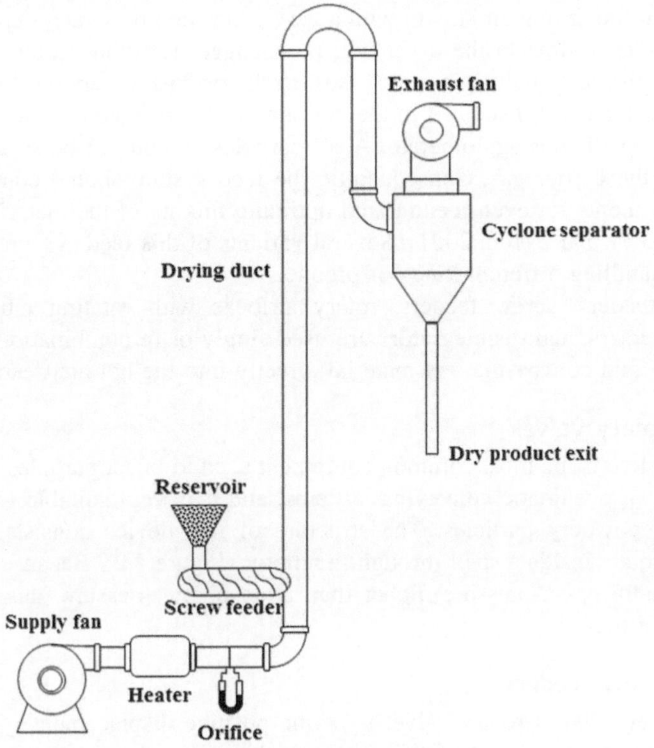

FIGURE 4.1 The schematic diagram of a simple flash dryer.

The heat source is typically installed directly in line with the circulated air. In some cases, heat is transferred to the circulated process air indirectly through an air-to-air heat exchanger.

Air (most common), inert gas (such as N_2 for drying solids wet with organic solvent), direct combustion gases, or superheated steam (or solvent vapor) can be used in convective drying systems as a drying medium. A direct heating system for heating the oil in the tank may be applied or an external one using a heat exchanger. The direct heating systems include bottom-fired gas strip burners, tubular heating systems, or electrical heating elements installed inside the tank.

The tubular heating systems transfer heat by passing burning gases, thermal oil, or steam through tubes that run through the vat from side to side. Electrical heating systems are convenient, but they are used mostly in batch dryers and small continuous systems due to higher energy costs. Electricity is used to heat-resistance elements inside small stainless-steel tubes assembled in grids.

4.4.3 Particle Feeding System

The short residence time (in seconds) in the pneumatic dryer requires a homogeneous feed material and very rapid, even dispersion of the wet material into the drying air stream. Moreover, the feeding mechanism should not supply a lumpy, wet solid into the drying air stream with a very controlled flow rate. Otherwise, the adjusted residence time in the dryer will be changed, resulting in the off specification of the product quality. In most cases, feeds for flash dryers must be granular and free flowing when dispersed in the gas stream. Such products neither stick on the conveyor walls nor agglomerate. Also, particles should not be so heavy as to drop out of the carrier gas. Consequently, the feed system should contain all the necessary elements for even feeding and optimum mixing of the materials and the drying air (Levy and Borde, 2001). Several variants of this feed system exist, each capable of handling different kinds of products.

Venturi feeders, screw feeders, rotary airlocks with rotating table feeders, mixers, dispersers, and disintegrators are used singly or in combination to control the feed rate and convey the wet material directly into the hot airstream.

4.4.3.1 Rotary Valves

The rotary valve is the most common equipment used to inject particles into the air passage line in pneumatic conveying systems, and it is very suitable for injecting free-flowing powdery particles. The structure of this device consists of several blades that rotate inside a shell through the motor (Figure 4.2). Because the drying gas line usually has a pressure higher than atmospheric pressure, this equipment must be an airlock.

4.4.3.2 Screw Feeders

This equipment, like a rotary valve, is of the positive displacement type, and by using the inverter, the speed of the spiral can be changed, and consequently, the injection particle flow rate can be changed. This method, like the rotary valve, can be used for positive and negative pressure lines. As the spiral rotates, the powder moves forward and enters the gas line through a duct located below or at the end of the screw (Figure 4.3). A screw feeder can be made with a variable diameter or stepped pitches depending on the type of application. It can also be used with several spirals inside the feeder.

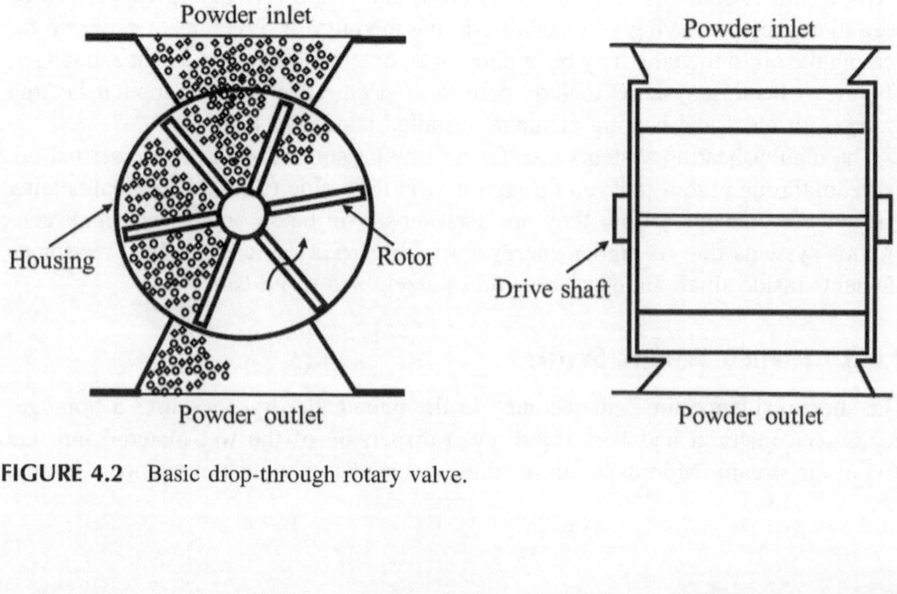

FIGURE 4.2 Basic drop-through rotary valve.

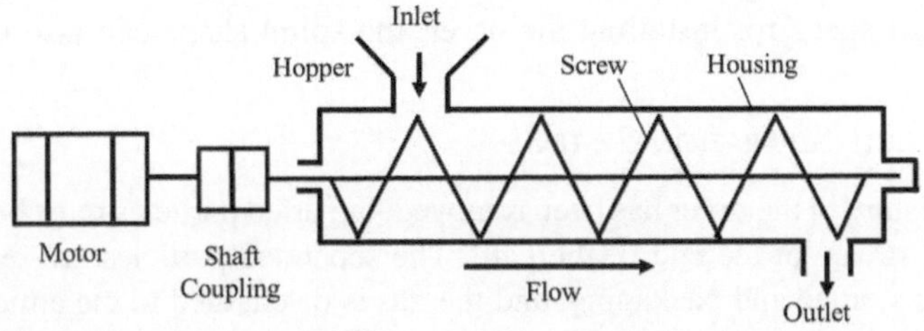

FIGURE 4.3 Simple screw feeder.

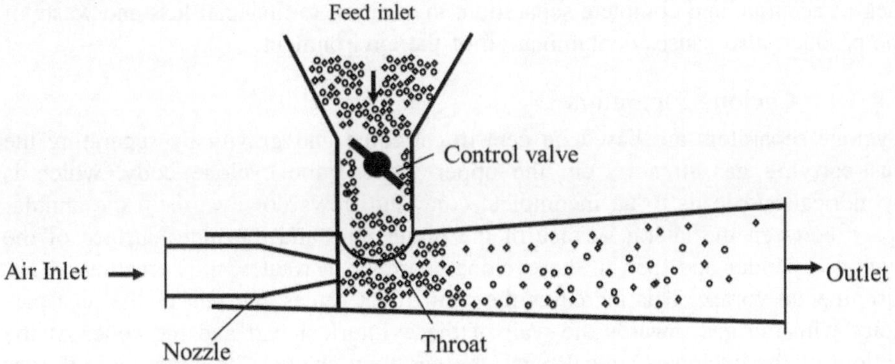

FIGURE 4.4 Commercial type of venturi feeder.

4.4.3.3 Venturi Feeders

Unlike rotary valves and screw feeders, which required an electric motor to transport particles, in this type of feeder, particles are drawn into the gas stream by the negative pressure created by the passage of gas flow through a nozzle. One of the limitations of this type of feeder is the material that should be free-flowing. It is also usually not possible to transfer high particle flow rates in long-length dryers. Because only low pressures can be used with the basic venturi, a standard industrial fan is often needed to supply the required air. Since the problem of negative pressure supply in this feeder is key, in the commercial type, which is in Figure 4.4, the design has changed somewhat. A flow control valve is also used to control the particle flow (Klinzing et al., 2011).

4.4.4 Drying Chamber

The pneumatic drying chamber is usually a circular tube with a specified diameter and length. Although it is possible to use the duct with other cross-sections shape, but it is not as common as a circular section. The length of the dryer depends on the residence time required to dry the particles. The dryer length must be sufficient to allow sufficient transfer of surface moisture to the drying gas. Pneumatic dryers are usually made as a straight tube, in which case it is also called a flash dryer. In the case of lack of space for installing the dryer, the spiral shape can also be used.

4.4.5 Particle Separation System

Once the moisture in the dryer has been removed, the dried particles must be separated from the gas stream at the end of the path. The separated particles are sent to other units, such as storage and packaging, and the gas is discharged to the atmosphere.

One of the main concerns of a pneumatic driving system designer is what separation system to use for gas-solid flow. The choice of separation system is a function of the particle size distribution. The importance of separation is that the lack of accurate and complete separation, in addition to financial loss and waste of the product, also causes contamination of the environment.

4.4.5.1 Cyclone Separators

Cyclone separators are based on centrifugal force and gravity by separating the dust-carrying gas stream from the upper wall of the cyclone body, which is cylindrical and leads to an incomplete cone and flows downward. In the annular space between the lateral surface of the outlet pipe and the inner surface of the cyclone cylinder and then in the cyclone chamber, it rotates, thus creating an environmental vortex. This increases the centrifugal forces and pushes the dust particles with the gas towards the wall of the cylindrical part and the cone. At the bottom of the cyclone (conical part), the gas flow changes its direction and goes from the center to the top and the outlet pipe. After contacting the cyclone wall and separating from the gas stream, the dust particles fall to the bottom of the cyclone by gravity and exit through the cyclone. The most common form of cyclone is the so-called reverse flow type, illustrated in Figure 4.5.

4.4.5.2 Filter

In pneumatic conveying systems handling fine or dusty material, the method of filtration that has become almost universally adopted is a bag-type fabric filter, either used on its own or as a backup to one of more cyclone separators. The gas–solid stream enters the device from beneath the fabric bags so that larger particles are separated by gravity settling, often aided by a cyclone action. However, this is not necessary, provided that direct impingement of particles on the bags from the conveying line is prevented. Fine particles are then caught on the insides of the fabric bags as the gas flows upward through the unit. The bags are usually of uniform cross-section along their length, and the most common shapes are circular or rectangular. Rectangular bags probably provide a filter unit with the largest fabric surface area to filter volume. The cleaning process is particularly important because it has a considerable influence on the size of filter required for a given application. Figure 4.6 diagrammatically illustrates a typical form of bag filter unit.

Although the filter bags shown in Figure 4.6 are suitable for continuously operating systems, the method of cleaning is only suitable for batch conveying operations, because filter surfaces cannot be cleaned effectively by shaking unless the flow of air ceases.

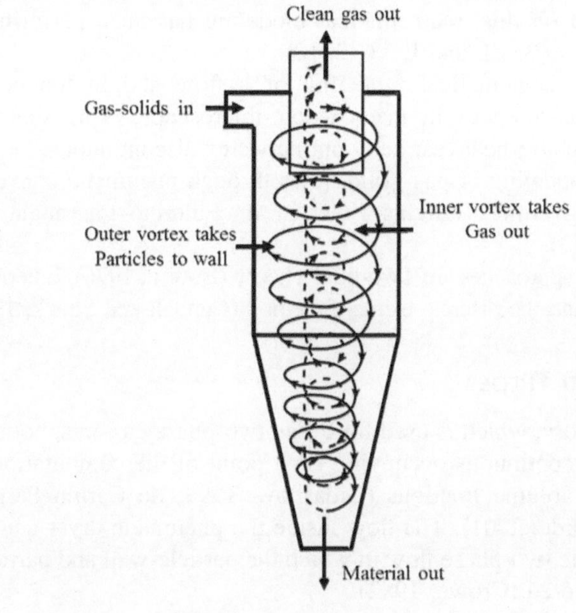

FIGURE 4.5 Reverse type cyclone separator.

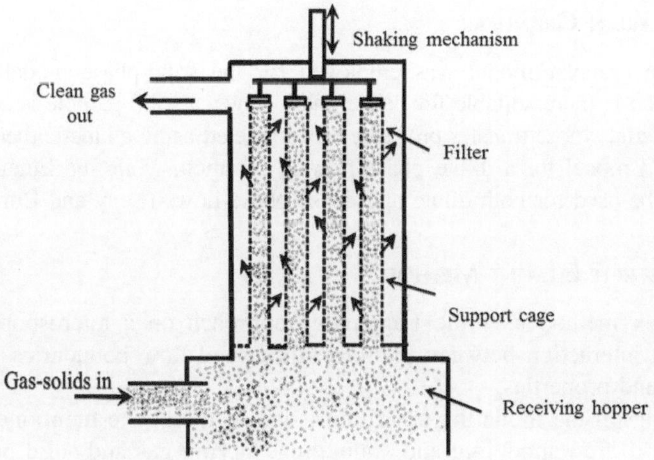

FIGURE 4.6 Typical shaken bag filter.

4.5 MATHEMATICAL MODELING

4.5.1 INTRODUCTION

To fully know the process, it is necessary to specify properties such as gas velocity and solid particles, temperature, pressure drop, and solid concentration. With this information, we can better understand the flow phenomena and better design the flow system. Based on this, mathematical modeling has been performed for different types of flows (Brosh and Levy, 2010).

The method of mathematical modeling of a flow and in the next step its numerical simulation refers to the macroscopic-microscopic expression of the flow and how the solid phase behaves; i.e., continuous or discontinuous.

Generally, the modeling of gas–solid flows through pneumatic conveying dryer can be done using either Eulerian–Eulerian or Eulerian–Lagrangian approach (Ibrahim et al., 2013).

These theoretical approaches are Two-fluid Theory (Bowen, 1976), Eulerian Granular (Gidaspow, 1994), and the discrete element method (Cundall and Strack, 1979).

4.5.2 Two-Fluid Theory

In the two-fluid theory, which is used for dilute two-phase currents, both phases are considered pseudo-continuous occupying each point of the computational domain according to their volume fractions (Gidaspow, 1994; do Carmo Ferreira et al., 2000; Levy and Borde, 2001). The flow inside the pneumatic dryer can usually be thought of as a dilute two-phase flow in which the particle-wall and particle-particle interactions are ignored (Crowe, 1982).

In two-fluid theory, the balance of mass, momentum, and energy equations are used macroscopic for both phases.

4.5.3 Eulerian Granular

The Eulerian granular model was employed for the solid-phase modeling. In this method, which is more suitable for dense phase flows, a solid particle is considered a granular material whose transfer properties are modeled using a kinetic theory method. The two-fluid model for a dilute phase flow is common, while the Eulerian granule method can be used for both dilute and dense phase flows (Levy and Borde, 2014).

4.5.4 Discrete Element Method

This complex method uses the Lagrangian approach on a microscopic scale to evaluate the interaction between particles, fluid, and flow boundaries using local parameters and properties.

Depending on the modeling conditions, one of the above-mentioned methods can be used to flow modeling and simulating. Drying gas and solid particles are traveling concurrently with the gas (Pelegrina and Crapiste, 2001; Tanaka et al., 2008).

4.6 EFFECT OF VARIOUS PARAMETERS APPLIED TO PNEUMATIC DRYING

There are different parameters that influence the drying capacity of a pneumatic dryer that affect the final moisture content of the product. These include the gas phase, dispersed phase, and geometrical properties, which are presented in Table 4.2.

TABLE 4.2
Effect of Various Parameters Applied to Pneumatic Drying

Category	Parameter	Description
Gas phase	Velocity	• As the inlet gas velocity increases, the particle transport velocity also increases, and the particle retention time will be shorter, resulting in a faster drying process (Namkung and Cho, 2004). • Increasing the gas velocity increases the mass and heat transfer coefficients between the gas and the particle and increases the equilibrium temperature between the two phases. This result is based on the constant particle flow and dryer diameter (Bunyawanichakul et al., 2007; Hidayat and Rasmuson, 2007). • Providing a faster gas flow means increasing the required hot gas flow rate, which requires more energy costs. • At a constant dryer length and gas temperature, increasing the gas velocity leads to an increase in the final moisture content of the product.
	Pressure drop	• Increasing the gas flow rate will reduce the pressure drop, while increasing the particle flow will increase the flow pressure drop. • Increasing the Reynolds number up to a certain Reynolds number, the pressure drop first decreases and then increases as the Reynolds number increases. The region of decreasing pressure is called the dense phase, and the rest is called the dilute phase. Particle static pressure is a parameter that determines the drop in flow pressure in the dense phase of pneumatic conveying systems (El-Behery et al., 2013).
	Temperature	• Increasing the temperature difference between the gas and the particle increases the drying force of the stream and causes moisture to separate from the particles and enter the gas stream. This transfer of moisture causes the particle temperature to increase and the gas temperature to decrease (Kemp, 1994). • Increasing the temperature drop of the gas along the drying path due to the effect on the gas velocity density causes a significant reduction in the gas velocity (Shigeru and Pei, 1984). • The higher inlet gas temperature increases the drying rate, but on the other hand, it may damage the quality of the product and cause cracks and discoloration (Kaensup et al., 2006).
	Humidity	• Increasing the moisture content of the gas leads to an increase in the moisture content of the final product (Shigeru and Pei, 1984). • Increasing the moisture content of the gas causes a slight decrease in the drying driving force. However, it has been shown that the temperature of a solid particle is affected by the gas humidity only during a constant drying rate period (Aubin et al., 2014).
	Reynolds number	• An increase in Reynolds number is usually caused by an increase in gas velocity, which reduces the particle residence time in the dryer and reduces the drying time. • The Reynolds number has a direct effect on the heat and mass transfer coefficients • Increasing the Reynolds number at a constant gas temperature means increasing the gas flow rate, which increases the amount of heat transferred to the fluid to evaporate the moisture from the particles.

(Continued)

TABLE 4.2 (*Continued*)
Effect of Various Parameters Applied to Pneumatic Drying

Category	Parameter	Description
Disperse phase	Diameter	• As the particle size increases, the contact surface of heat transfer between the gas and the particle decreases. This reduction in heat transfer increases the particle temperature (El-Behery et al., 2013). • Larger particles will have a higher final moisture content due to the lower moisture transfer with the hot gas (Paixao and Rocha, 1998; El-Behery et al., 2013). • Smaller particles increase the moisture content of the gas due to the higher moisture transfer with the hot gas (El-Behery et al., 2013).
	Shape	• Particles with higher lateral surface-to-volume ratios than spheres have higher mass and heat transfer coefficients (Pelegrina and Crapiste, 2001).
	Flow rate	• As the gas-solid ratio increases, the particle drying rate increases, and the final moisture content of the particles will decrease remarkably (Lv et al., 2012). • As the particle flow rate increases, the slip velocity will increase, which will improve the displacement convective transport coefficient (Pelegrina and Crapiste, 2001).
Geometrical	Dryer diameter	• Increasing the diameter of the dryer will reduce the velocity of the gas and the particle and will also reduce the slip velocity between the two phases. As a result, mass and heat transfer coefficients will be reduced (Paixao and Rocha, 1998). • Increasing the diameter of the dryer reduces the particle residence time. Therefore, the contact time between the two phases and the mass and heat transfer between the phases is reduced, and the particles lose less moisture. Eventually, the final moisture content of the particles will be higher as they leave the dryer (Paixao and Rocha, 1998). • The final equilibrium temperature of the two phases increases with increasing dryer diameter. This factor causes the particle temperature and moisture diffusion rate to rise. As a result, the content of the final moisture content will be lower (Bunyawanichakul et al., 2007). • Once the system has reached equilibrium temperature, increasing the dryer diameter will reduce the final moisture content of the particle, and the gas moisture content will also increase. Therefore, increasing the dryer diameter can reduce the drying capacity at a constant particle flow rate and constant inlet gas velocity. Unlike the final particle moisture, the gas moisture and equilibrium temperature increase with increasing dryer diameter (Bunyawanichakul et al., 2007).
	Dryer height	• The particles must be present in the dryer for a specific minimum residence time, which means the dryer must have a minimum length. If the length of the dryer is less than the specified value, specific energy consumption will increase sharply. • Increasing the dryer length increases the retention time of the particles and their heat and moisture exchange with the gas. Thus, the final moisture content of the particles is reduced as an essential parameter (Kemp, 1994).

4.7 RECENT DEVELOPMENTS

The improvement of performance and product quality and drying time and energy consumption reduction are the most important reasons for making changes to a system.

The use of combined systems is one of the ways to improve drying systems. Using a flash dryer as a pre-dryer for other systems such as fluidized bed dryer (batch or continuous), fluidized bed cooler, spray dryer, and drum dryer can improve process efficiency. Research by Dorfeshan et al. (Dorfeshan et al., 2022) showed that the flash-fluidized bed combination can significantly improve system performance. In this combination system, the flash dryer has the task of drying the wet cake and assassinates surface moisture, and the bound moisture is separated in the fluidized bed dryer.

The use of pneumatic dryers as a second step after a packed bed has been investigated. The results show that it has a positive effect in increasing the drying capacity and reducing energy consumption (Frodeson et al., 2013). The combination of flash and cyclone dryers is also used, especially in drying polymeric materials such as PVC suspension. In some studies, it has been introduced to be more efficient than the combination of flash-rotary (Korn, 2007).

Impinging stream drying is a novel alternative to pneumatic drying at high drying loads. In this method, the intensive collision of opposed streams creates a zone that makes high heat, mass, and momentum transfer and rapid surface moisture removal (Choicharoen et al., 2010). Very useful information about the ISD system is provided by Mujumdar (Mujumdar, 2014). Moisture reduction depends on both the impinging steam drying temperature and tempering time in the air impinging stream dryer (Pruengam et al., 2014). The results of a study on the combined impinging stream and pneumatic dryer showed that the oscillatory motion of particles can extended the residence time and amplify the drying process (Nimmol et al., 2012). Also, an increase in the impinging distance leads to the improved drying performance of an ISD.

Multistage drying, superheated steam drying, and improvement of the agitated flash dryer are methods that can be further developed and use their capabilities. Another type of flash dryer that has both flash and fluid-bed properties is the spin flash dryer, which is suitable for applications with high solid flow rates (Levy and Borde, 1999).

REFERENCES

Aubin, A.; Ansart, V.; Hemati, M.; Lasuye, T.; Branly, M. Modeling and simulation of drying operations in PVC powder production line: Experimental and theoretical study of drying kinetics on particle scale. *Powder Technology* 2014, 255, 120–133.

Baeyens, J.; Van Gauwbergen, D.; Vinckier, I. Pneumatic drying: The use of large-scale experimental data in a design procedure. *Powder technology* 1995, 83(2), 139–148.

Bowen, R. M. Theory of mixtures. *Continuum physics* 1976, 3(1).

Brosh, T.; Levy, A. Modeling of heat transfer in pneumatic conveyer using a combined DEM-CFD numerical code. *Drying Technology* 2010, 28(2), 155–164.

Bunyawanichakul, P.; Walker, G. J.; Sargison, J. E.; Doe, P. E. Modelling and simulation of paddy grain (rice) drying in a simple pneumatic dryer. *Biosystems Engineering* 2007, 96(3), 335–344.

Choicharoen, K.; Devahastin, S.; Soponronnarit, S. Performance and energy consumption of an impinging stream dryer for high-moisture particulate materials. *Drying Technology* 2010, 28(1–3), 20–29.

Crapiste, G. H.; Rotstein, E. Sorptional equilibrium at changing temperatures. In: A. S. Mujumdar (Ed.), *Drying of Solids. Recent International Developments*, New Delhi: Wiley Eastern, 1986, 41–45.

Crowe, C. T. Review—Numerical models for dilute gas-particle flows. *Journal of Fluids Engineering* 1982, 104(3), 297–303.

Cundall, P. A.; Strack, O. D. A discrete numerical model for granular assemblies. *Geotechnique* 1979, 29(1), 47–65.

Debrand S. Heat transfer during a flash drying process. *Industrial & Engineering Chemistry Process Design and Development* 1974, 13(4), 396–404.

Dorfeshan, M., Mehrzad, S., Hajidavalloo, E. Experimental and numerical investigation of specific energy consumption of the two-stage pneumatic-fluidized bed drying of suspension-grade polyvinyl chloride. *Drying Technology* 2022, 40, 371–386.

do Carmo Ferreira, M.; Freire, J. T.; Massarani, G. Homogeneous hydraulic and pneumatic conveying of solid particles. *Powder Technology* 2000, 108(1), 46–54.

El-Behery, S. M.; El-Askary, W. A.; Hamed, M. H.; Ibrahim, K. A. Numerical simulation of heat and mass transfer in pneumatic conveying dryer. *Computers & Fluids* 2012, 68, 159–167.

El-Behery, S. M.; El-Askary, W. A.; Hamed, M. H.; Ibrahim, K. A. Eulerian–lagrangian simulation and experimental validation of pneumatic conveying dryer. *Drying Technology* 2013, 31(12), 1374–1387.

Frodeson, S.; Berghel, J.; Renström, R. The potential of using two-step drying techniques for improving energy efficiency and increasing drying capacity in fuel pellet industries. *Drying Technology*, 2013, 31(15), 1863–1870.

Gidaspow, D. *Multiphase Flow and Fluidization: Continuum and Kinetic Theory Descriptions*. Academic Press; San Diego, California, 1994.

Hidayat, M.; Rasmuson, A. Heat and mass transfer in U-bend of a pneumatic conveying dryer. *Chemical Engineering Research and Design* 2007, 85(3), 307–319.

Ibrahim, K. A.; Hamed, M. H.; El-Askary, W. A.; El-Behery, S. M. Swirling gas–solid flow through pneumatic conveying dryer. *Powder technology* 2013, 235, 500–515.

Indarto, A.; Halim, Y.; Partoputro, P. Pneumatic drying of solid particle: Experimental and model comparison. *Experimental Heat Transfe*r 2007, 20(4), 277–287.

Kaensup, W.; Kulwong, S.; Wongwises, S. Comparison of drying kinetics of paddy using a pneumatic conveying dryer with and without a cyclone. *Drying Technology* 2006, 24(8), 1039–1045.

Kemp, C. I. Scale-up of pneumatic conveying dryers. *Drying Technology* 1994, 12(1–2), 279–297.

Klinzing, George E., et al. *Pneumatic Conveying of Solids: A Theoretical and Practical Approach.* Vol. 8. Springer Science & Business Media; New York, 2011.

Korn, O. Cyclone Dryer: A pneumatic dryer with increased solid residence time. *Drying Technology* 2007, 19(8), 37–41.

Levy, A.; Borde, I. Steady state one-dimensional flow model for a pneumatic dryer. *Chemical Engineering and Processing: Process Intensification* 1999, 38(2), 121–130.

Levy, A.; Borde, I. Two-fluid model for pneumatic drying of particulate materials. *Drying Technology* 2001, 19(8), 1773–1788.

Levy, A.; Borde, I. *Pneumatic and Flash Drying. In Handbook of Industrial Drying*, 4th Ed.; Boca Raton, FL: CRC Press, 2014, 381–392.

Lv, W.; Li, R. Y.; Wang, X. H.; Li, Y. D.; Sun, H. W. The application of mathematical model and numerical simulation in the pneumatic drying of the straw. *Advanced Materials Research* 2012, 354, 290–293.

Mezhericher, M.; Levy, A.; Borde, I. Three-dimensional modelling of pneumatic drying process. *Powder Technology* 2010, 203(2), 371–383.

Mujumdar, A. S. Impingement drying. In *Handbook of Industrial Drying*, 4th Ed.; Boca Raton, FL: CRC Press, 2014, 371–379.

Namkung, W.; Cho, M. Pneumatic drying of iron ore particles in a vertical tube. *Drying Technology* 2004, 22(4), 877–891.

Nimmol, C.; Sathapornprasath, K.; Devahastin, S. Drying of high-moisture paddy using a combined impinging stream and pneumatic drying system. *Drying Technology* 2012, 30(16), 1854–1862.

Paixao, A. E.; Rocha, S. C. Pneumatic drying in diluted phase: Parametric analysis of tube diameter and mean particle diameter. *Drying technology* 1998, 16(9–10), 1957–1970.

Pelegrina, A. H.; Crapiste, G. H. Modelling the pneumatic drying of food particles. *Journal of Food Engineering* 2001, 48(4), 301–310.

Pruengam, P.; Soponronnarit, S.; Prachayawarakorn, S.; Devahastin, S. Rapid drying of parboiled paddy using hot air impinging stream dryer. *Drying Technology*, 2014, 32(16), 1949–1955.

Shigeru, M.; Pei, D. C. A mathematical analysis of pneumatic drying of grains—I. Constant drying rate. *International journal of heat and mass transfer* 1984, 27(6), 843–849.

Skuratovsky, I.; Levy, A.; Borde, I. Two-dimensional numerical simulations of the pneumatic drying in vertical pipes. *Chemical Engineering and Processing: Process Intensification* 2005, 44(2), 187–192.

Tanaka, F.; Uchino, T.; Hamanaka, D.; Atungulu, G. G. Mathematical modeling of pneumatic drying of rice powder. *Journal of Food Engineering* 2008, 88(4), 492–498.

5 Drying of Particulates by Impinging Streams

Somkiat Prachayawarakorn
Department of Chemical Engineering, Faculty of Engineering, King Mongkut's University of Technology Thonburi, Bangkok, Thailand

Sakamon Devahastin
Department of Food Engineering, Faculty of Engineering, King Mongkut's University of Technology Thonburi, Bangkok, Thailand

The Academy of Science, The Royal Society of Thailand, Bangkok, Thailand

Somchart Soponronnarit
Division of Energy Technology, School of Energy, Environment and Materials, King Mongkut's University of Technology Thonburi, Bangkok, Thailand

The Academy of Science, The Royal Society of Thailand, Bangkok, Thailand

CONTENTS

DOI: 10.1201/9781003207108-5

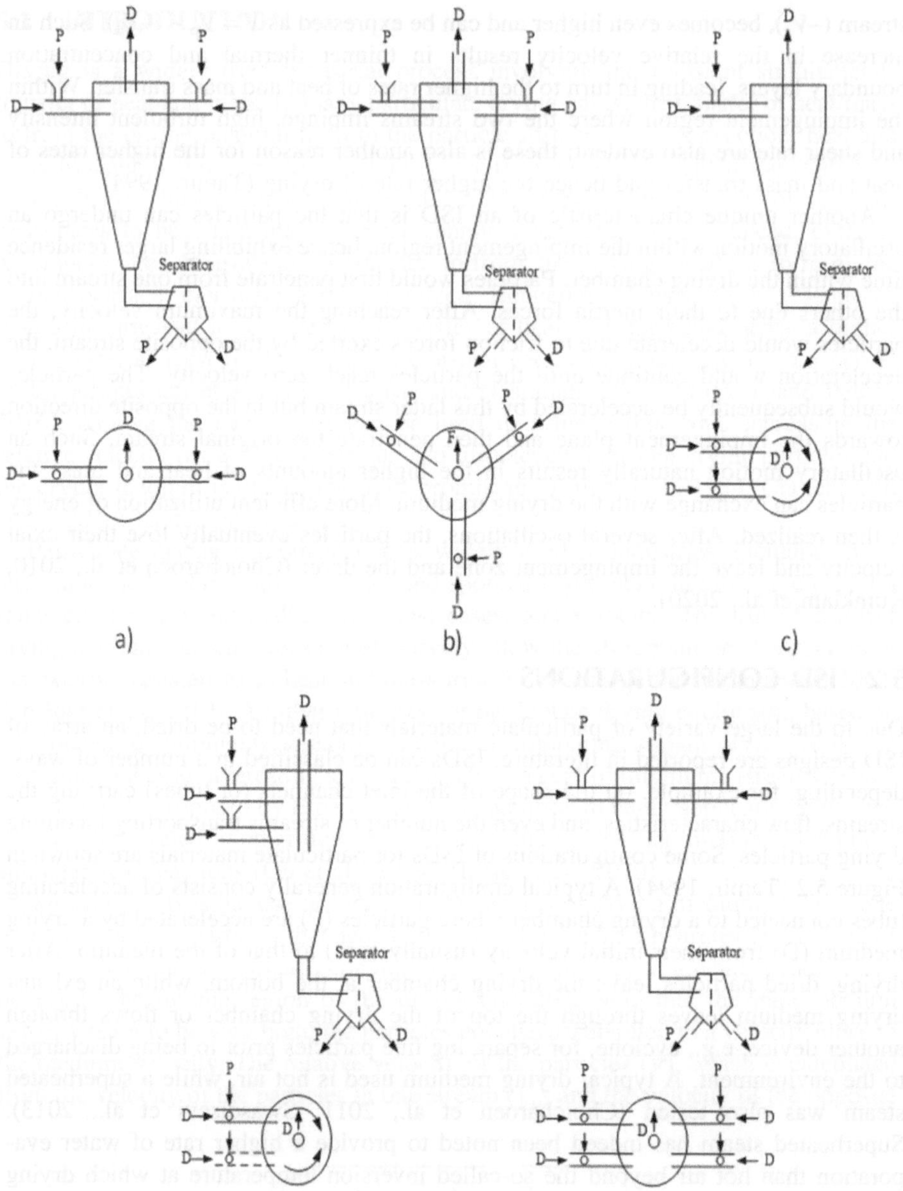

FIGURE 5.2 ISDs of various configurations (Tamir, 1994). (a) Coaxial horizontal two impinging streams; (b) horizontal three impinging streams; (c) curvilinear two impinging streams; (d) curvilinear four impinging streams; (e) four impigning streams. D = drying medium; P = particles.

- Eccentric countercurrent: more than two streams of drying medium flow on the different axes. As shown in Figure 5.2b, three streams, each containing drying particles, impinge at the middle plane within the drying chamber.
- Coaxial curvilinear: two streams of drying medium, each containing drying particles, flow tangentially in a countercurrent direction, as shown in Figures 5.2c and 5.2d. The streams flow on the wall of the drying chamber along the streamlines represented by half circle before they collide with each other.
- Non-coaxial curvilinear: two streams of drying medium tangentially and countercurrently enter the drying chamber, with central lines in different planes before collision, as shown in Figure 5.2e. The streams flow on the wall of the drying chamber along the streamlines represented by several circles.

ii. Impingment region
- Stationary: impingement region (or plane) is fixed in position.
- Moving: impingement region (or plane) periodically or continuously moves.

5.3 ADVANTAGES AND LIMITATIONS OF ISD

ISD belongs to a class of rapid drying equipment, similar to a flash dryer. ISD exhibits high water evaporation capacity and is therefore very competitive when compared with other types of particulate dryers, e.g., moving bed, fluidized bed, and spouted bed dryers. Impinging streams can also help break agglomerated particles, rendering drying of sticky materials more effective (Choicharoen et al., 2010). Due to its higher effectiveness, from a heat and mass transfer point of view, an ISD can be designed to have a more compact size at the same required drying load. Its design is also simple due to lack of moving parts.

Despite its many advantages, ISD possesses some limitations. ISD requires higher electric energy consumption due to the need of a larger fan/blower to deliver high-velocity impinging streams of drying medium; pressure drop within its mixed drying medium and particulate piping system is also naturally larger (Kumklam et al., 2020). Erosion due to constant contacts between high-velocity moving particles and piping system may also require attention (Aderibigbe and Rajaratnam, 1996). It is also important to note that although an ISD may result in a very high water evaporation rate and hence a shorter required drying time in the unhindered drying rate period, removal of intra-particle water during the falling rate period remains a challenge. For this reason, while an ISD may serve as an attractive alternative for surface moisture removal, bound moisture removal may not be suitably conducted in this type of dryer. ISD may indeed serve as a first-stage dryer to remove surface water of particles before being sent to another second-stage dryer that allows drying to take place over a longer period of time (Nimmol et al., 2012).

A comparison of performance in terms of water evaporation capacity, drying time, and air requirement between a coaxial ISD and fluidized bed dryer (FBD) for parboiled paddy drying are given in Table 5.1 (Bootkote et al., 2016; Kumklam et al., 2020).

TABLE 5.1
Comparison of Performance Between Coaxial ISD and Fluidized Bed Dryer for Parboiled Paddy Drying (Bootkote et al., 2016; Kumklam et al., 2020)

Performance	ISD	FBD*
Moisture evaporation capacity (kg water/m^3·h)	2044	216
Drying time (s)	23	120
Air mass flow rate(kg air/kg evaporated water)	2.3	13.6

Note
* FBD = Fluidized bed dryer

The initial and final moisture contents of a paddy were in the ranges of 47%–55% (d.b.) and of 22%–25% (d.b.), respectively. The tested ISD exhibited ninefold higher moisture evaporation capacity and sixfold shorter required drying time than the tested fluidized bed. Such a higher evaporataion capacity and shorter required drying time come with the higher required air mass flow rate, however.

Table 5.2 gives a comparison between the performance of an ISD and that of selected spray dryer designs (Kudra and Mujumdar, 1989). The ZT-100, 1–3.2/18 BK, and NEMO are the common spray dryers with different volumes. The listed operating conditions are almost similar; the inlet and outlet drying temperatures were in the ranges of 140°C–150°C and 50°C–70°C. The comparison is therefore reasonable. The evaporation rates are noted to be similar between the ISD and spray dryers. However, the volumetric evaporation capacity of the ISD is seen to be significantly higher, indicating that the required volume of the ISD is much smaller at a similar required drying capacity (or load).

In addition to its advantage in terms of higher drying rate, an ISD also results in some other benefits from a product quality point of view. When being applied to the drying of a parboiled paddy, for example, the vigorous impingement or, in other words, collision of air/particle streams results in the splitting of rice husks from the endosperm (Kumklam et al., 2020). The coaxial ISD used in such an investigation is schematically shown in Figure 5.3, wherein one stream of air carrying a parboiled paddy is impinged with the opposite stream. Since a rice husk needs to be removed from dried parboiled rice kernels by a dehusking machine prior to milling, spontaneous splitting of the husk implies a reduction of at least one step during the milling process. In addition, water can be more rapidly removed from the huskless endosperm as the husk acts as a moisture barrier. Drying time, which is naturally related to the production cost, can therefore be reduced even further. More details on this benefit of an ISD will again be discussed in a later section.

Another application of an ISD involves particle collision for simultaneous drying and grinding of a particulate material. A schematic diagram of the ISD dryer-grinder is shown in Figure 5.4 (Elperin, 1972). In this case, a particulate material is fed into an airlift tube and allowed to flow through a separator and classified into different fractions of different sizes. Large and wet portions of the particles are

TABLE 5.2
Comparison of Performance of ISD and Selected Spray Dryer Designs (Kudra and Mujumdar, 1989)

Dryer Type (Industrial)	Evaporation Rate (kg Water/h)	Evaporative Capacity (kg Water/m^3·h)	Air Temperature		Dimension		Volume (m^3)	Reference
			Inlet temp. (°C)	Outlet temp. (°C)	Length (m)	Diameter (m)		
ZT-100 (GDR)	100	3.3	140	50	3.5	3.3	30	Unpublished data
	100	5.5	140	70	2.2	3.2	18	
1–3.2/18 BK (USSR)	80	3.7	150	70	–	2.9	22	Luikov (1970)
"NEMO" (FRG)	80	2.5	140	70	–	2.9	32	Luikov (1970)
ISD (USSR)	100	28.0	150	70	4.0	1.2	4.4	ITMO an BSSR

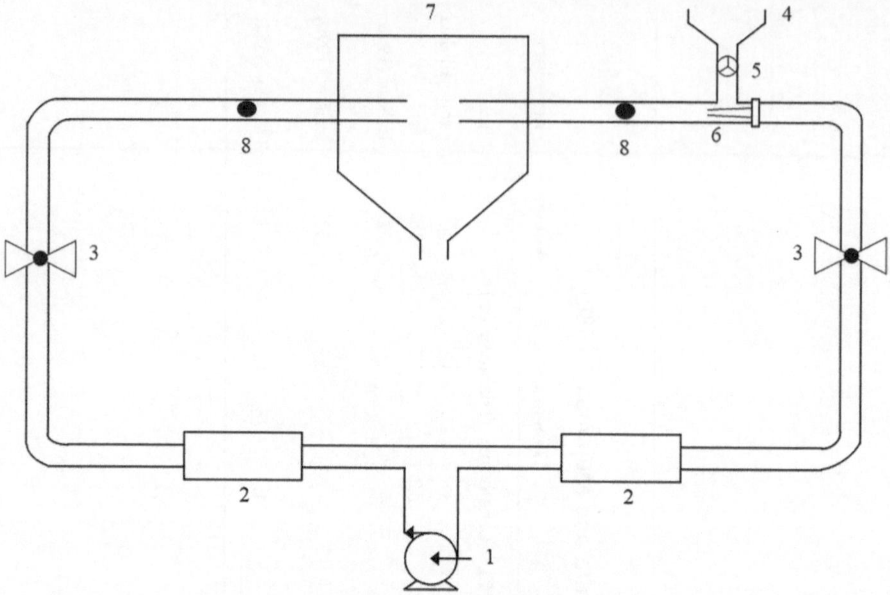

FIGURE 5.3 Schematic diagram of ISD used for parboiled paddy drying. (1) High-pressure blower; (2) electric heaters; (3) globe valves; (4) feed hopper; (5) star feeder; (6) nozzle; (7) drying chamber; (8) inlet air temperature measurement ports (Kumklam et al., 2020).

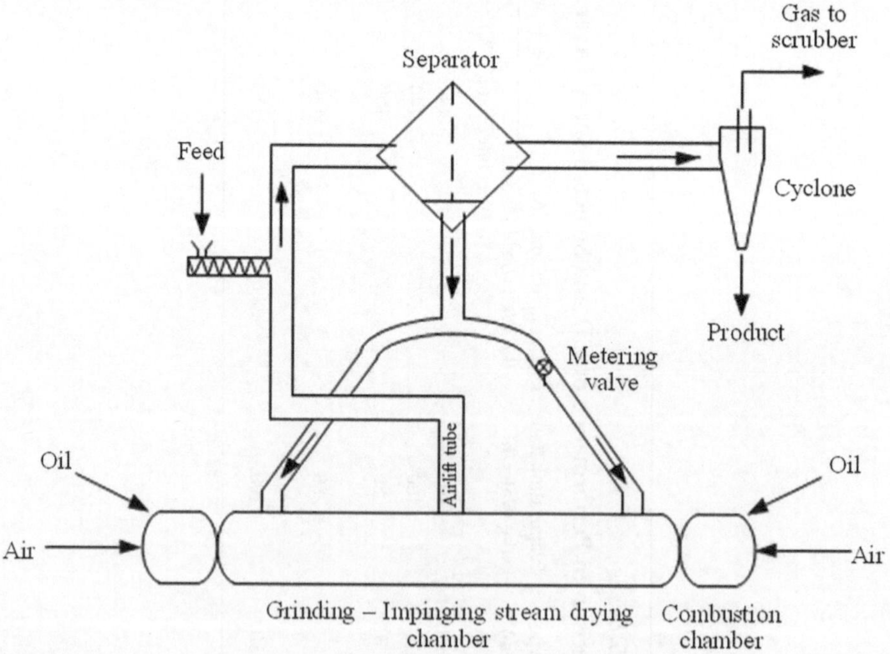

FIGURE 5.4 Schematic diagram of impinging stream drying-grinding chamber for a particulate material (Elperin, 1972).

moved into a drying-grinding chamber, where drying and grinding simultaneously take place; grinding expectedly takes place due to particle collisions. Exhaust air carries over fine particles through a cyclone; air leaving the cyclone is cleaned by a scrubber prior to being released into the atmosphere. Since grinding proceeds rapidly, overheating of particles can be avoided. This type of dryer is therefore applicable for drying-grinding a heat-sensitive particulate material.

As mentioned earlier, an ISD exhibits some limitations that hamper its industrial application. The most obvious limitation is related to its shorter residence time of particles in the dryer, only approximatey 1–2 s (Pruengam et al., 2014; Kumklam et al., 2020). Such a short residence time would not allow adequate removal of water from solid particles. Particles, especially agricultural or food particles, cannot be dried to a safe storage moisture content in a single drying stage. Note that an arrangement of multiple ISDs is relatively difficult, making multi-stage impractical for industrial applications. Lengthening of particle residence time is therefore a primary concern that needs attention in future designs and developments of an ISD.

5.4 PRACTICAL APPLICATIONS OF ISD

5.4.1 Drying of a Parboiled Paddy

A parboiled paddy is a paddy that has partially or fully been cooked prior to drying. Three major steps are indeed involved in the production of parboiled paddy, namely, soaking, steaming (or cooking), and drying. A parboiled paddy after steaming typically has a moisture content in the range of 40%–45% (d.b.), which is significantly higher than that of a fresh paddy (Tirawanichakul et al., 2004; Pruengam et al., 2014). From a practical point of view, a fluidized bed, circulating batch, or an LSU (Louisiana State University) dryer can be used to dry a parboiled paddy.

Due to the high surface moisture nature of a parboiled paddy and to the limited efficiency of the competing dryers, an ISD is a good candidate for parboiled paddy drying. A coaxial ISD, with one stream containing a parboiled paddy is shown in Figure 5.3, and was indeed tested for its performance in terms of moisture removal capability, energy consumption, and ability to maintain head rice yield of the dried paddy (Kumklam et al., 2020). Figure 5.5 shows the effects of various relevant operating parameters, including the feed rate, superficial air velocity, inlet drying air temperature, and impinging distance on the moisture content reduction of a parboiled paddy. The experiments were carried out at the drying temperatures of 150°C–190°C, feed rates of 80–160 $kg_{dry\ solid}$/h, impinging distances of 5–13 cm, and superficial air velocities of 10–25 m/s. The drying cycle mentioned in the figure refers to the number of times a parboiled paddy passed through and was dried by the ISD.

At less than five drying cycles, the moisture content slightly decreased and all the operating parameters did not affect the moisture reduction despite the high turbulence within the drying chamber. This is because the moisture content of a rice husk was only around 21.5% (d.b.); this value was significantly lower than that of the rice endosperm, with a moisture content of around 60% (d.b.). The internal moisture needed time to travel to the surface of the paddy. This fact was also confirmed by

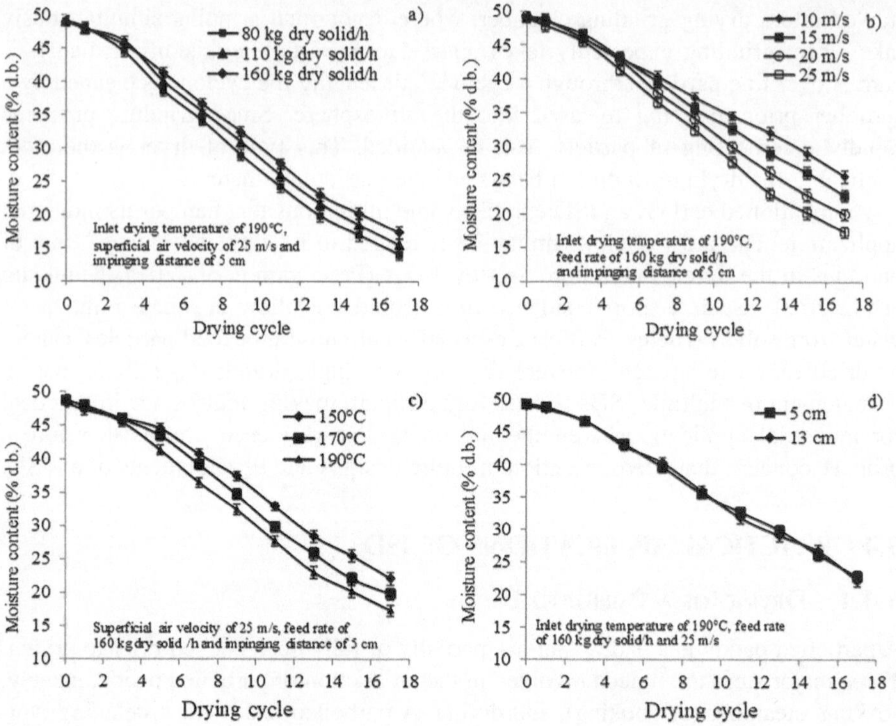

FIGURE 5.5 Effects of (a) feed rate, (b) superficial air velocity, (c) inlet drying air temperature, and (d) impinging distance on moisture reduction of parboiled paddy undergoing impinging stream drying (Kumklam et al., 2020).

the large moisture gradients within the husk during the early drying cycles of less than five (see Figure 5.6). The moisture gradients became smaller as the number of drying cycles increased (Thuwapanichayanan et al., 2022). The smaller moisture gradients, especially in the rice husk, beneficially resulted in higher drying rates; the moisture was more readily available for removal on the surface of the paddy. The high level of turbulence in the ISD could now enhance the moisture evaporation rate, as can be seen in Figure 5.6 when the number of drying cycles was larger than five. Higher drying air temperature, air velocity, and lower feed rate also resulted in higher drying rates. Impinging distance nevertheless did not significantly affect the drying rate since the shorter impinging distance might not noticeably increase the turbulence intensity. Hence, it did not facilitate the transfer of moisture from the paddy surface to the bulk airstream.

One important aspect that is affected by the vigorous movement and collision of the streams as well as particles in the ISD is the splitting of the husk from the endosperm. Figure 5.7 shows the amount of dehusked kernels at various drying conditions. The percentage of dehusked kernels increased with an increasing drying cycle, indicating expectedly that more repeated collisions resulted in the larger amount of the dehusked kernels. Drying air temperature and impinging distance did

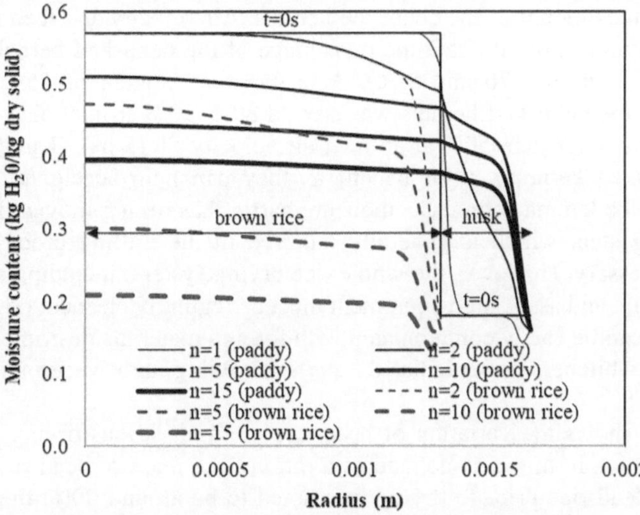

FIGURE 5.6 Simulated moisture content distributions inside paddy and brown rice kernels undergoing impinging stream drying for different drying cycles. Inlet air temperature = 190°C, superfical air velocity = 25 m/s, feed rate = 160 kg dry solid/h (Thuwapanichayanan et al., 2022).

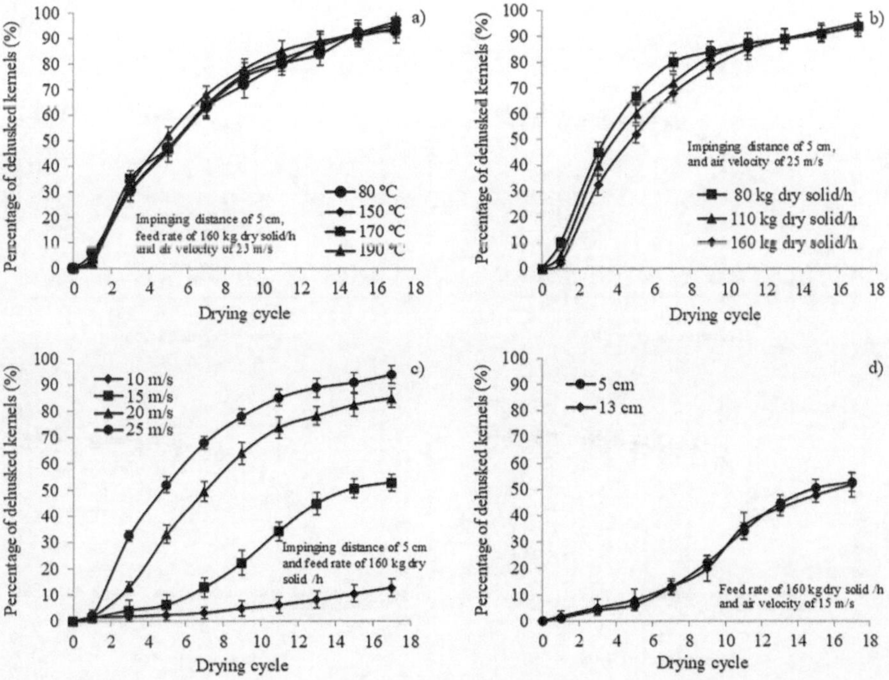

FIGURE 5.7 Changes in percentage of dehusked kernels as a function of operating conditions of ISD. (a) Inlet drying air temperature; (b) feed rate; (c) superficial air velocity; and (d) impinging distance (Kumklam et al., 2020).

not influence the amount of the dehusked kernels. Air velocity and feed rate, on the other hand, significantly affected the percentage of the dehusked kernels, again as expected (see Figures 5.7b and 5.7c). At a moisture content of 25% (d.b.), the percentage of the dehusked kernels was around 80%–90% at inlet air velocities of 20–25 m/s and lower than 50% at an inlet air velocity of 15 m/s (Kumklam et al., 2020). Dehusked kernels are of benefit as they can help accelerate the drying process since the lemma acting as a moisture barrier has been removed. In addition, the dehusking step, which is generally required in the milling process, may no longer be necessary. However, the whole rice drying system, including a conveyer, a tempering bin, and associated equipment must be regularly cleaned; otherwise, the final product can be easily contaminated with foreign materials or grime. Failure of a polisher or whitener, in particular the parts involving abrasive grinding wheels, may also result.

Figure 5.8 shows the variation of head rice yield at various drying conditions. Head rice yield is in this case defined as a ratio of the mass of head rice to that of brown rice. Head rice prior to drying was noted to be around 90%; the remaining quantity was rice bran. Inlet drying air temperature, air velocity, feed rate, and impinging distance did not affect the head rice yield. The moisture content of a parboiled paddy after drying was observed to be the only factor that caused the reduction in the head rice yield. More rapid reduction in the head rice yield was evident as the moisture content of parboiled paddy after drying became lower than

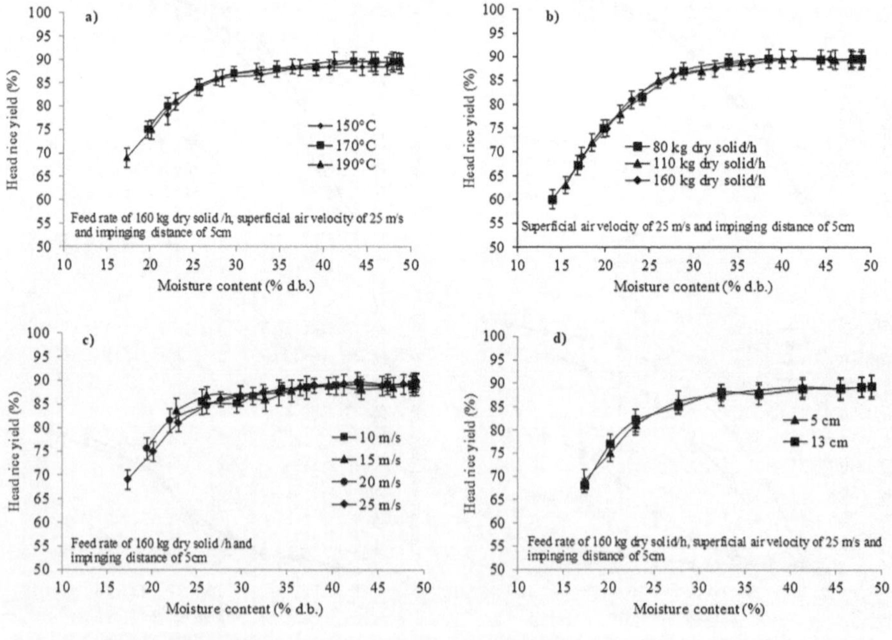

FIGURE 5.8 Changes in head rice yield as a function of operating conditions of ISD. (a) Inlet drying air temperature; (b) feed rate; (c) superficial air velocity; and (d) impinging distance (Kumklam et al., 2020).

25% (d.b.). Such a moisture content value is indeed noted as a limited final moisture content for drying parboiled paddy in an ISD. At this moisture content, the head rice yield was around 85%. It is important and interesting to note, however, that accelerating the drying process did not adversely affect the head rice yield. Therefore, increasing the parboiled paddy drying capacity can easily be achieved by enhancing the drying rate. In the case of a non-parboiled paddy, on the other hand, excessive drying rate would result in the reduction of head rice. This is because the mechanical properties of parboiled paddy, including tensile and compressive strengths, are significantly more superior to those of a non-parboiled paddy (Thakur and Gupta, 2006; Wetchakama et al., 2019). The insignificant influences of the drying conditions on the head rice yield at moisture contents higher than 25% (d.b.) are probably due to the superior mechanical properties, particularly tensile strength, of a parboiled paddy to those of a non-parboiled paddy (Wetchakama et al., 2019).

Figure 5.9 shows the values of the total specific energy consumption, including thermal and electric energy consumptions, of the ISD at various drying conditions. An increase in both the inlet drying air temperature and feed rate could save more energy. The total specific energy consumption was around 14 MJ/kg evaporated water when drying was conducted at 190°C at an air velocity of 25 m/s and feed rate of 160 $kg_{dry\ solid}$/h, with no exhaust air recycle and a cyclone dust collector. When a cyclone was connected to the drying chamber, the mean residence time of a paddy increased by 70% (relative to that in the system with no cyclone); the corresponding energy consumption reduced to 8.6 MJ/kg evaporated water (Thuwapanichayanan et al., 2022). These results indicate that lengthening the mean residence time of particles results in a more effective energy utilization within an ISD.

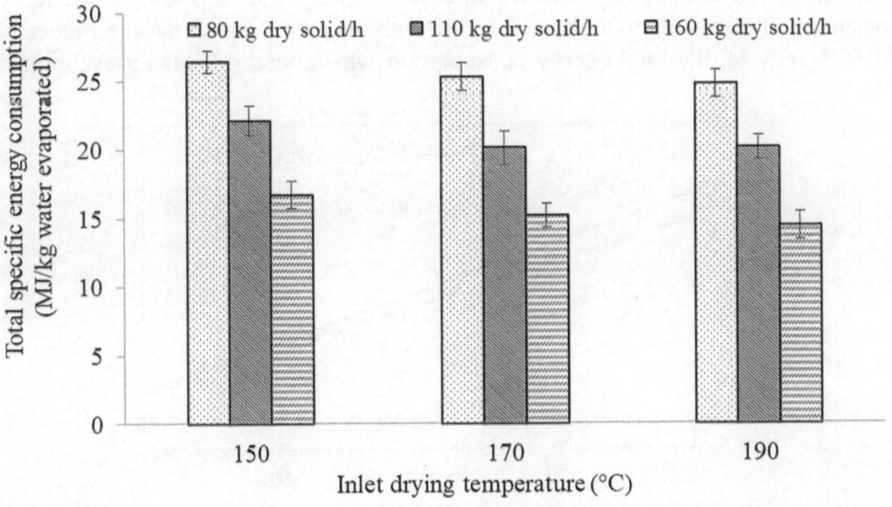

FIGURE 5.9 Variation of total specific energy consumption at different operating conditions of ISD processing parboiled paddy. Impinging distance = 5 cm, superficial air velocity = 25 m/s and intermediate moisture content = 25% (d.b.) (Kumklam et al., 2020).

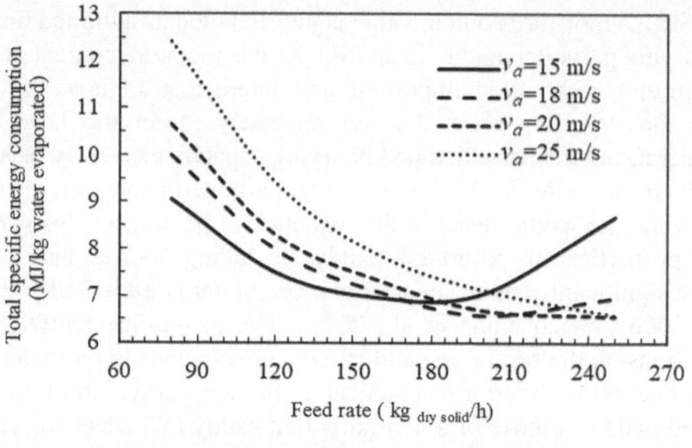

FIGURE 5.10 Simulated effects of superficial air velocity and feed rate on total specific energy consumption of ISD processing parboiled paddy. Inlet drying air temperature = 190°C with no recycle of exhaust air and a cyclone (Thuwapanichayanan et al., 2022).

The effects of inlet air velocity, feed rate, and percentage air recycle on the total specific energy consumption in a coaxial ISD were also simulated (Thuwapanichayanan et al., 2022). As can be seen in Figure 5.10, the air velocity and feed rate within the ranges of study affected the total specific energy consumption. When suitable air velocities and feed rates were selected, the energy consumption would not be much different, being in the range of 7–8 MJ/kg evaporated water (when the ISD was operated without exhaust air recycle).

Figure 5.11 depicts the effect of exhaust air recycle on the total specific energy consumption at the suitably selected inlet air velocity and feed rate. The energy consumption was noted to linearly decrease with the percentage exhaust air recycle. At 80% recycle, the total energy consumption was around 4.0 MJ/kg evaporated

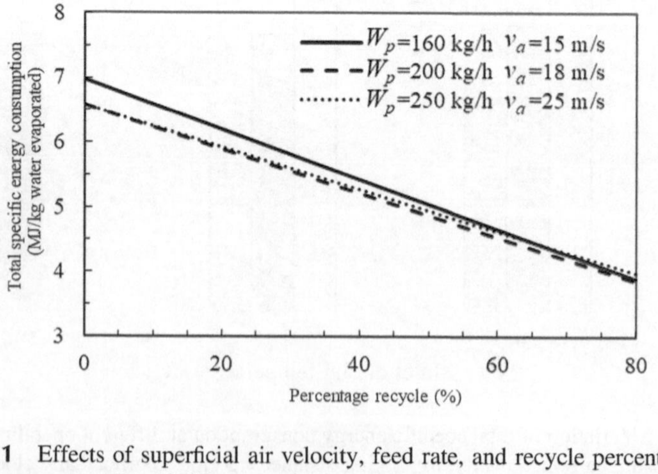

FIGURE 5.11 Effects of superficial air velocity, feed rate, and recycle percentage on total specific energy consumption of ISD processing parboiled paddy. Inlet drying air temperature = 190°C (Thuwapanichayanan et al., 2022).

water when reducing the moisture content of paddy from 45% to 25% (d.b.) at an inlet drying air temperature of 190°C. Such an energy consumption value was lower than those belonging to other typical grain dryers such as a fluidized bed, spouted bed, and inclined bed dryers, which exhibit the specific total energy consumption of around 6–8 MJ/kg evaporated water (Madhiyanon and Soponronnarit, 2005; Prachayawarakorn et al., 2005; Sarker et al., 2013). These results indicate that while the energy consumption of an ISD with no exhaust air recycle may not be different from those of other competing dryers, adding an air recycle step can lead to a significant reduction in the energy consumption of an ISD. In addition, the effective drying time, which is defined as the time that kernels are exposed to the drying air, when reducing the moisture content from 50% to 25% (d.b.) was only 0.5 min in the case of the tested ISD; other competing dryers require 4 min or longer for the same drying task.

From the dryer performance point of view, an ISD can be regarded as a highly efficient dryer. However, when applying to a parboiled paddy drying, multi-stage dryers need to be installed should a safe storage moisture content is to be reached. Such multi-stage arrangement requires good planning and management. To be more practical, the required number of drying cycles to reach a desired moisture content should be as much as possible reduced. A means to lengthen the time particles spend in an ISD viz. particle residence time is therefore highly desirable and should be further investigated.

5.4.2 Enhancement of Bioactive Compounds in a Germinated Paddy

Germination is a process of growth of plants. In the case of a cereal seed, during germination, the seed would absorb water; the embryo then develops and produces phytohormones, especially gibberellin. Such hormones would diffuse into the aleurone layer via the scutellum. The aleurone layer cells are in turn induced to produce hydrolytic enzymes. The endosperm, mainly consisting of starch, is hydrolyzed from simple sugars and proteins by the enzymes into peptides and amino acids (Ayernor and Ocloo, 2007). Among various enzymes, gamma-aminobutyric acid (GABA), which is a non-proteinaceous amino acid, is commonly found in a germinated paddy (Srisang et al., 2011; Chungcharoen et al., 2014). This amino acid serves as a predominant inhibitory neurotransmitter in the central nervous system. GABA is effective in decreasing the blood pressure of animals and humans (Inoue et al., 2003), inhibiting cancer cell proliferation (Oh and Oh, 2004), and even reducing the risk of Alzheimer's disease and many forms of allergies (Nakamura et al., 2004).

Enhancement of GABA in a germinated paddy can be done either by chemical or physical means. Chemically, a paddy can be soaked in water at near-neutral pH (pH 6) to obtain a higher GABA content; such a content is nevertheless not statistically different from that obtained via soaking at pH 4 or 7 (Choi et al., 2004). Oh and Choi (2000) alternatively reported that brown rice germinated for 72 h in a chitosan solution (1,000 ppm) exhibited 1.3-fold higher GABA content than the control sample germinated in distilled water. Through a physical means, which is preferable by health-conscious consumers, a soaked paddy germinated under hypoxic condition has

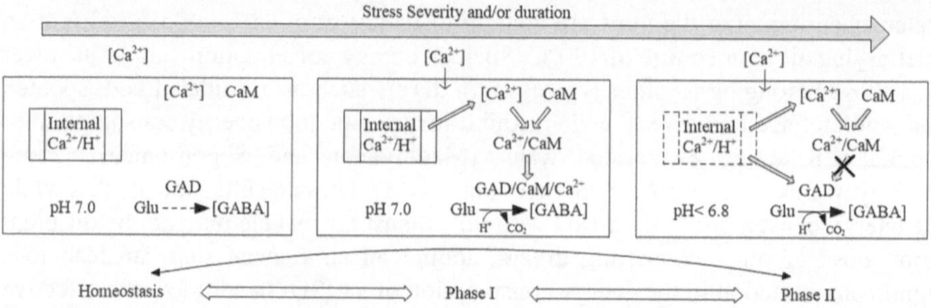

FIGURE 5.12 Biphasic regulation of GAD activity. At physiological pH, stress-induced increase in cellular Ca^{2+} complex with CaM and GAD is activated by Ca^{2+}/CaM. Intracellular storage of Ca^{2+} into the cytosol further amplifies the stress response. Below pH 6.8, acid activates GAD with a pH optima at 5.8, at which GAD activity is 3–9 times higher than that at 7.3 in the presence of Ca^{2+}/CaM. Stress-induced membrane damage releases H+ from vacuoles into the cytosol (Kinnersley and Turano, 2000).

been noted to exhibit higher GABA content. Komatsuzaki et al. (2007), for example, reported that the GABA content of brown rice germinated under hypoxic condition was 0.3 to 2.5-fold higher than that of the rice germinated by a conventional soaking method. In the former case, brown rice was soaked in water at 35°C for 24 h and kept under hypoxic condition at 35°C for 21 h. Such enhancement is ascribed to the fact that glutamic acid is produced via the glutamate synthase-glutamine synthetase pathway, which plays an important role in the anaerobic accumulation of GABA and alanine (Reggina et al., 2000).

Heat shock is another physical method that can be used to enhance the production of GABA in a germinated paddy. Heat shock stimulates plant stress defence mechanisms, which in turn leads to the production of plant secondary metabolites, including GABA. The biphasic stimulation of glutamate decarboxylase (GAD) and control of GABA in plants exposed to a variety of stresses such as cold and heat shocks is schematically illustrated in Figure 5.12. For the sake of simplicity, the stimulation of GAD activity is divided into two phases based on Ca^{2+} and calmodulin-dependent (CAM) (Phase I) or acidic pH-dependent (Phase II) activity.

The stimulation of GAD by Ca^{2+}/CaM in Phase I may serve as a rapid response to stress and/or a response to a mild and transient stress (Sanders et al., 1999). As cytosolic pH decreases due to the extended duration and/or severity of stress, GAD activity could then be stimulated by acidic pH in a Ca^{2+}/CaM-independent manner in Phase II (Kinnersley and Turano, 2000). If stress is transient and/or mild, however, Phase II may not be reached and the cells would revert to normal metabolism directly from Phase I. Alternatively, under severe conditions, Phase II may be the only type of control (pH-dependent) of GAD activity.

ISD has proved to be an excellent means for heat-shock pretreatment due to its ability to rapidly deliver the energy to a paddy. Techo et al. (2018) indeed employed a coaxial ISD to a heat shock-soaked paddy prior to germination. The results are shown in Figure 5.13 for three paddy varieties, namely, SP1, KDML105, and KD-MJ2. Note that the amylose contents of SP1, KDML105, and KD- MJ2 were 29%, 20%, and 6%,

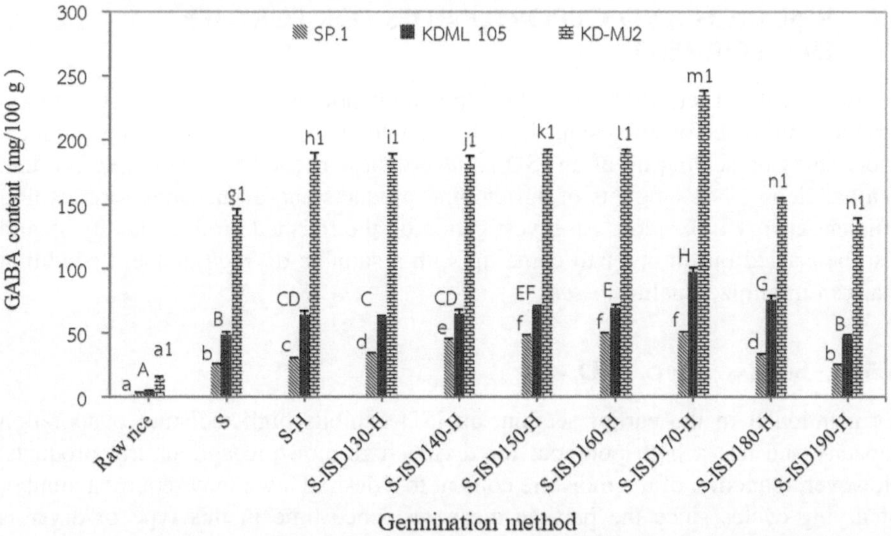

FIGURE 5.13 Effects of germination method and heating temperature on GABA content of three paddy varieties (Techo et al., 2018). S=Soaking; S-H=Soaking and hypoxia, S-ISD130-H=Soaking and impinging steam at 130°C followed by hypoxia; S-ISD140-H= Soaking and impinging steam at 140°C followed by hypoxia; S-ISD150-H= Soaking and impinging steam at 150°C followed by hypoxia; S-ISD160-H= Soaking and impinging steam at 160°C followed by hypoxia; S-ISD170-H= Soaking and impinging steam at 170°C followed by hypoxia; S-ISD180-H= Soaking and impinging steam at 180°C followed by hypoxia; S-ISD190-H= Soaking and impinging steam at 190°C followed by hypoxia.

respectively. GABA contents of the raw paddy of SP1, KDML105, and KD-MJ2 varieties were 2.88, 5.05, and 13.57 mg/100 g, respectively. The paddy that was soaked and then was followed by a heat treatment and hypoxia (S-ISD-H) or the one that was soaked and followed by hypoxia (S-H) exhibited higher GABA contents than that undergone only soaking (S), with S-ISD-H samples exhibiting the highest contents. The higher GABA content is ascribed to the aforementioned mechanisms.

As shown in Figure 5.13, the heat treatment temperature positively and negatively affects the GABA content. For example, in the case of SP1, an increase in the temperature up to 170°C resulted in the higher GABA content; beyond this temperature, the GABA content significantly decreased. Appropriate treatment temperatures were not different for the three paddy varieties; the suitable treatment temperature was 170°C for SP1, KDML105, and KD-MJ2. This is because this treatment (or inlet air) temperature would result in the grain temperature at the dryer outlet of approximately 40°C, which is the suitable temperature for the GAD enzymatic activity, as reported by Zhang et al. (2007). At the suitable treatment temperature, the GABA contents were 49.4, 97.8, and 235 mg/100 g for SP1, KDML105, and KD-MJ2 varieties, respectively. These results indicate that KD-MJ2 is the most sensitive variety to heat shock; this is followed by KDML105 and SP1. The appropriate particle mean residence time was noted to be in the range of 1–2 s.

5.5 RESEARCH AND OPPORTUNITIES FOR FURTHER DEVELOPMENT

As discussed earlier, ISD has a high potential not only for drying agricultural products but also for enhancing bioactive compounds contents in some produce. More study on scaling up of an ISD is nevertheless needed to obtain a design that can handle massive amounts of agricultural products and at the same time exhibit efficient-energy utilization. An investigation on the finished product quality should also be carried out in order to come up with a suitable drying/treatment condition that can minimize quality losses.

5.5.1 Scaling Up of ISD

As mentioned in the earlier section, an ISD exhibits high moisture evaporation capacity and has a high potential for drying high-moisture agricultural products. However, reduction of the moisture content to a desired level may require a number of drying cycles since the particle mean residence time in this type of dryer is generally very short. Multiple drying cycles are nevertheless not convenient in practice. Extension of the particle mean residence time is therefore needed to reduce the required number of drying cycles. Design and scaling up of an ISD are unfortunately very complicated and require a thorough knowledge of both continuous-phase (i.e., drying medium) and disperse-phase (i.e., particles) transport phenomena (Choicharoen et al., 2012). Trajectories and other behaviors of the drying particles need to be adequately predicted (Choicharoen et al., 2012). Note that particle trajectory is heavily influenced both by the configuration and dimensions as well as the operating conditions (e.g., inlet air velocity and particle feed rate) of a dryer. Experimental study alone would be impossible to yield complete information of the aforementioned phenomena (Kumklam et al., 2020).

Several computational studies have indeed been conducted to investigate the effects of relevant factors on transport phenomena and hence performance of various ISD designs. Among various computational models, computational fluid dynamics in combination with discrete element method (CFD-DEM) has proved to adequately simulate the multiphase transport phenomena within an ISD (Khomwachirakul et al., 2016). An appropriate heat and mass transfer model should also be included in order to simulate the transport processes within drying particles during drying. This is of importance as not only surface moisture needs to be removed but also intra-particle moisture. An ability to predict changes in the moisture and temperature, which is closely intertwined with the moisture, is therefore required. CFD-DEM model has a potential to be used for the design of a pilot-scale ISD where particle-particle and particle-fluid interactions are complicated.

5.5.2 Quality of Dried Products

Quality of a dried product is a primary parameter that can be used as a criterion to assess an industrial dryer, in addition to its drying capacity and energy consumption. Although much research has been devoted to the study of ISD performance, quality of dried products after impinging stream drying has rarely been documented. This is a gap that must be filled should an ISD is to be developed into a more viable drying technology for the food industry. Dried product quality can be considered in such terms as nutritional value, color, and texture. It is hypothesized that the quality of a dried product obtained via this drying technique should be higher as the time a product is exposed to the drying air is very short, even when multiple drying cycles are required. However, an ISD may not be applicable to all agricultural products, in particular those that exhibit weak structure since they may easily deform or be damaged due to the extensive collisions within the impingement zone.

An area that may be suitable for an ISD involves seed germination or sprouting. As mentioned in the earlier section, GABA content in germinated rice can be enhanced by the use of an ISD, where rapid heat transfer can selectively enhance the formation of some chemical constituents in the biphasic stimulation of GAD to produce a series of chemical reactions and hence more GABA. Such an enhancement method is purely physical in nature; it should be of great interest to the food and even biotechnological industries seeking a chemical-free method to enhance health-beneficial compounds in various natural products. Nevertheless, the extent of enhancement depends on the variety and employed heating temperature as mentioned earlier. Further investigation should be conducted on other products, including wheat and millet seeds as well as various beans.

ISD also has a potential to puff grains or some fruits for the production of breakfast cereals and snacks. Due to its rapid heat and mass transfer capability, intra-particle moisture would rapidly vaporize. Such internal vaporization and hence vapor generation would in turn lead to hydrostatic pressure buildup, structural expansion, and eventually puffing of a product. A highly porous structure that forms as a result of puffing is expected to lead to a crisp texture, which is highly desirable in the case of snacks. Rehydration capability is also expected to significantly increase as a result of void formation. Such products can be produced without frying, so they should serve the demand of today's health-conscious consumers.

5.6 CONCLUDING REMARKS

Despite its potential for drying high-moisture particulate materials, an impinging stream dryer (ISD) still needs to gain more momentum and acceptance from relevant industries. More research is clearly required to reach this goal. One major challenge that must be tackled is the limited particle residence time in the dryer; a means to increase the residence time is needed if a multi-stage dryer set up, which may not be too convenient in practice, is to be avoided. The shorter particle mean residence would be beneficial; on the other hand, if an ISD is to be used for the high-temperature short time stimulation of some chemical reactions, e.g., the activation of glutamate decarboxylase enzyme to enhance the formation of gamma-aminobutyric acid (GABA) in germinating grains. It is important to note that appropriate operating conditions must be identified and utilized to achieve such benefits.

To enhance its industrial feasibility and usability, scaling up of an ISD is an important task. Due to the complications of particle-particle and particle-fluid interactions within an ISD, computational fluid dynamics in combination with discrete element method (CFD-DEM) should be used to simulate the multiphase transport phenomena; the simulated information should then be used for scaling up the dryer. The quality of dried particulate materials, especially food and other biomaterials, also needs much further investigation.

ACKNOWLEDGMENTS

The authors express their sincere appreciation to the Thailand Research Fund (Grant no. DPG5980004 and RGU6180004), Program Management Unit for Human Resources & Institutional Development, Research and Innovation (PMU-B) (Grant no. B05F640155) and the National Science and Technology Development Agency (NSTDA; Grant no. P-20-52263) for their financial supports, leading to some of the results included in this chapter.

REFERENCES

Aderibigbe, O. O. and Rajaratnam, N. 1996. Erosion of loose beds by submerged circular impinging verticle turbulent jets. *Journal of Hydraulic Research* 34: 19–33.

Ayernor, G. S. and Ocloo, F. C. K. 2007. Physico-chemical changes and diastatic activities associated with germinating paddy rice (PSB.Rc 34). *African Journal of Food Science Research* 3: 203–207.

Bootkote, P., Soponronnarit, S. and Prachayawarakorn, S. 2016. Process of producing parboiled rice with different colors by fluidized bed drying technique including tempering. *Food and Bioprocess Technology* 9: 1574–1586.

Choi, H.-D., Park, Y.-K., Kim, Y.-S., Chung, C.-H. and Park, Y.-D. 2004. Effect of pretreatment conditions on γ-aminobutyric acid content of brown rice and germinated brown rice. *Korean Journal of Food Science and Technology* 36: 761–764 (in Korean).

Choicharoen, K., Devahastin, S. and Soponronnarit, S. 2010. Performance and energy consumption of an impinging stream dryer for high-moisture particulate materials. *Drying Technology* 28: 20–29.

Choicharoen, K., Devahastin, S. and Soponronnarit, S. 2011. Comparative evaluation of performance and energy consumption of hot air and superheated steam impinging stream dryers for high-moisture particulate materials. *Applied Thermal Engineering* 31: 3444–3452.

Choicharoen, K., Devahastin, S. and Soponronnarit, S. 2012. Numerical simulation of multiphase transport phenomena during impinging stream drying of a particulate material. *Drying Technology* 30: 1227–1237.

Chungcharoen, T., Prachayawarakorn, S., Soponronnarit, S. and Tungtrakul, P. 2014. Effects of germination process and drying temperature on gamma-aminobutyric acid (GABA) and starch digestibility of germinated brown rice. *Drying Technology* 32: 742–753.

Elperin, I. T. 1961. Heat and mass transfer in impinging streams. *Inzhenerno Fizicheski Zuhr* 6: 62–68.

Elperin, I. T. 1972. *Transport Processes in Opposing Jets (Gas Suspensions)*. Science and Technology Press, Minsk (in Russian).

Gu, R., Li, X., Cheng, K. and Wen, L. 2019. Application of micro-impinging stream reactors in the preparation of Co and Al Co-Doped $Ni(OH)_2$ nanocomposites for supercapacitors and their modification with reduced graphene oxide. *RSC Advances* 9: 25677–25689.

Inoue, K., Shirai, T., Ochiai, H., Kasao, M., Hayakawa, K., Kimura, M. and Sansawa, H. 2003. Blood-pressure-lowering effect of a novel fermented milk containing γ-aminobutyric acid (GABA) in mild hypertensives. *European Journal of Clinical Nutrition* 57: 490–495.

Kaensup, W. and Wongwises, S. 2004. Combined microwave/fluidized bed drying of fresh peppercorns. *Drying Technology* 22: 779–794.

Khomwachirakul, P., Devahastin, S., Swasdisevi, T. and Soponronnarit, S. 2016, Simulation of flow and drying characteristics of high-moisture particles in an impinging stream dryer via CFD-DEM. *Drying Technology* 34: 403–419.

Kinnersley, A. M. and Turano, F. J. 2000. Gamma aminobutyric acid (GABA) and plant responses to stress. *Critical Reviews in Plant Sciences* 19: 479–509.

Komatsuzaki, N., Tsukahara, K., Toyoshima, H., Suzuki, T., Shimizu, N. and Kimura, T. 2007. Effect of soaking and gaseous treatment on GABA content in germinated brown rice. *Journal of Food Engineering* 78: 556–560.

Kudra, T. and Mujumdar, A. S. 1989. Impingement stream dryers for particulates and pastes. *Drying Technology* 7: 219–266.

Kudra, T. and Mujumdar, A. S. 2015, Special drying techniques and novel dryers, *In Handbook of Industrial Drying, 4th ed.* A. S. Mujumdar (Ed.), 434–487. CRC Press, Boca Raton.

Kumklam, P., Prachayawarakorn, S., Devahastin, S. and Soponronnarit, S. 2020. Effects of operating parameters of impinging stream dryer on parboiled rice quality and energy consumption. *Drying Technology* 38: 634–645.

Luikov, A. 1970. Drying in Chemical Industry. Khimia, Moscow (in Russian).

Madhiyanon, T. and Soponronnarit, S. 2005. High temperature spouted bed paddy drying with varied downcomer air flows and moisture contents: Effects on drying kinetics, critical moisture content, and milling quality. *Drying Technology* 23: 473–495.

Mohseni, M. and Peters, B. 2016. Effects of particle size distribution on drying characteristics in a drum by XDEM: A case study. *Chemical Engineering Science* 152: 689–698.

Najjari, M. and Nasrallah, S. B. 2009. Heat transfer between a porous layer and a forced flow: Influence of layer thickness. *Drying Technology* 27: 336–343.

Nakamura, K., Tian, S. and Kayahara, H. 2004. Functionality enhancement in germinated brown rice. *Proceedings of the 11th International Flavor Conference/3rd George Charalambous Memorial Symposium*. Samos, Greece, 356–371.

Nimmol, C., Sathapornprasath, K. and Devahastin, S. 2012. Drying of high-moisture paddy using a combined impinging stream and pneumatic drying system. *Drying Technology* 30: 1854–1862.

Oh, C.-H. and Oh, S.-H. 2004. Effect of germinated brown rice extracts with enhanced levels of GABA on cancer cell proliferation and apoptosis. *Journal of Medicinal Food* 7: 19–23.

Oh, S.-H. and Choi, W.-G. 2000. Production of the quality germinated brown rice containing high γ-aminobutyric acid by chitosan application. *Korean Journal of Biotechnology and Bioengineering* 15: 615–620 (in Korean).

Ottosson, A., Nilsson, L. and Berghel, J. 2017. A mathematical model of heat and mass transfer in yankee drying of tissue. *Drying Technology* 35: 323–334.

Prachayawarakorn, S., Tia, W., Poopaiboon, K. and Soponronnarit, S. 2005. Comparison of performances of pulsed and conventional fluidised-bed dryers. *Journal of Stored Products Research* 41: 479–497.

Pruengam, P., Soponronnarit, S., Prachayawarakorn, S. and Devahastin, S. 2014. Rapid drying of parboiled paddy using hot air impinging stream dryer. *Drying Technology* 32: 1949–1955.

Reggina, R., Nebuloni, M. and Brambilla, I. 2000. Anaerobic accumulation of amino acids in rice roots: Role of the glutamine synthetase/glutamate synthase cycle. *Amino Acids* 18: 207–217.

San José, M. J., Alvarez, S. and López, R. 2019. Drying of industrial sludge waste in a conical spouted bed dryer: Effect of air temperature and air velocity. *Drying Technology* 37: 118–128.

Sanders, D., Brownlee, C. and Harper, J. F. 1999. Communicating with calcium. *The Plant Cell* 11: 691–706.

Sarker, M. S. H., Ibrahim, M. N., Aziz, N. A. and Punan, M. S. 2013. Drying kinetics, energy consumption, and quality of paddy (MAR-219) during drying by the industrial inclined bed dryer with or without the fluidized bed dryer. *Drying Technology* 31: 286–294.

Sathapornprasath, K., Devahastin, S. and Soponronnarit, S. 2007. Performance evaluation of an impinging stream dryer for particulate materials. *Drying Technology* 25: 1111–1118.

Srisang, N., Varanyanond, W., Soponronnarit, S. and Prachayawarakorn, S. 2011. Effect of heating media and operating conditions on drying kinetics and quality of germinated brown rice. *Journal of Food Engineering* 107: 385–392.

Swasdisevi, S., Devahastin, S., Thanasookprasert, S. and Soponronnarit, S. 2013. Comparative evaluation of hot air and superheated-steam impinging stream drying as novel alternatives for paddy drying. *Drying Technology* 31: 717–725.

Tamir, A. 1994. *Impinging-Stream Reactors*, Elsevier, Amsterdam.

Tamir, A., Elperin, I. and Luzzatto, K. 1984. Drying in a new two impinging streams reactor. *Chemical Engineering Science* 39: 139–146.

Techo, J., Soponronnarit, S., Prachayawarakorn, S. and Devahastin, S. 2018. Effect of thermal treatment during rice germination on γ-aminobutyric acid, antioxidant activity and texture of germinated rice of different varieties. *KMUTT Research and Development Journal* 41: 371–379 (in Thai).

Thakur, A. K. and Gupta, A. K. 2006. Two stage drying of high moisture paddy with intervening rest period. *Energy Conversion and Management* 47: 3069–3083.

Thuwapanichayanan, R., Kumklam, P., Soponronnarit, S. and Prachayawarakorn, S. 2022. Mathematical model and energy utilization evaluation of a coaxial impinging stream drying system for parboiled paddy. *Drying Technology* 40: 158–174.

Tirawanichakul, S., Prachayawarakorn, S., Varanyanond, W., Tungtrakul, P. and Soponronnarit, S. 2004. Effect of fluidized bed drying temperature on various quality attributes of rice. *Drying Technology* 22: 1731–1754.

Wetchakama, S., Prachayawarakorn, S. and Soponronnarit, S. 2019. Change of mechanical properties related to starch gelatinization and moisture content of rice kernel during fluidized bed drying. *Drying Technology* 37: 1173–1183.

Zhang, H., Yao, H.-Y., Chen, F. and Wang, X. 2007. Purification and characterization of glutamate decarboxylase from rice germ. *Food Chemistry* 101: 1670–1676.

Zhang, J-W, Li, W.-F., Xu, X.-L., El-Hassan, M., Liu, H.-F. and Wang, F. C. 2020. Effect of geometry on engulfment flow regime in T-jet reactors. *Chemical Engineering Journal* 387: 124148.

6 Drying of Suspensions and Pastes in Spouted Beds

José Teixeira Freire and Maria do Carmo Ferreira
Department of Chemical Engineering, Federal University of São Carlos, São Carlos, SP, Brazil

Fábio Bentes Freire
Department of Chemical Engineering, Federal University of São Carlos, São Carlos, SP, Brazil

Department of Civil Engineering, Federal University of Technology, Curitiba, PR, Brazil

Flavio Bentes Freire
Department of Civil Engineering, Federal University of Technology – Paraná, Curitiba, Brazil

Ronaldo Correia de Brito
Department of Chemical Engineering, Federal University of São Carlos, São Carlos, SP, Brazil

CONTENTS

6.1 INTRODUCTION

Many industrial processes concerned with powder production rely on thermal drying to remove water or solvents from intermediate products in the form of pastes and suspensions. Spray-drying remains the most important technique for large-scale processing and industrial applications aimed at drying liquid suspensions. Despite being widely used, spray-dryers have high installation costs and might have limitations for processing products with determined features, such as too viscous or too sticky materials. Atomizing or dripping the product in an agitated vessel containing particles suspended in a hot air flow emerged as an alternative technique to the spray drying. It takes advantage of the high turbulence and inter-particle contact that are characteristic features of fluidized, spouted, and vibrofluidized systems. The first studies on drying organic dyes and dye intermediates in spouted beds date back to

DOI: 10.1201/9781003207108-6

the 1960s (Mathur and Epstein, 1974) and, since then, agitated beds with inert particles have been extensively investigated to dry a broad variety of suspensions and pasty-like materials. The spouted bed dryer in particular has many attractive features for this application, such as the low installation costs, the flexibility to operate under a wide range of operational conditions, the possibility of enhancing heat and mass transfer by performing simple alterations in the geometrical configurations. At the same time, modeling and simulation of paste drying in SBs are quite challenging, owing to the complex forces developed during the interaction of the liquid, gas, and solid phases in the drying process.

This chapter presents recent advances achieved in drying with inert particles, focusing mainly on the studies investigating the spouting technique with inert particles published in the last decade. A compilation of research aimed at investigating this technique to process diverse materials and offering solutions to improve the operational stability and the quality attributes of the finished powders is presented. The heat and mass transfer mechanisms that rule the drying process are briefly approached, as well as the advances in modelling and process control. Modified configurations of agitated beds developed to process sticky or difficult-to-handle products are discussed and the industrial uses and future perspectives for academic research in this field are addressed.

6.2 SPOUTING TECHNIQUE WITH INERT PARTICLES

Pasty materials can be obtained either mixing a liquid with a powder or by crushing and mixing different substances. Many raw and input materials used in different industrial sectors, as well as several by-products and residues that come out from processing are pastes. Often, these wet materials are thermally dehydrated to build a powdery product for further use or for disposal.

A few decades ago, when drying over inert particles was still a promising paste drying technique, Strumillo et al. (1983) proposed a concise classification for the pasty materials. They considered the following groups: hard pastes (filter cakes, precipitates); soft pastes and sludges (starch dispersions, pulps); suspensions (lime, colloidal gold), emulsions (raw or skimmed milk), and solutions (animal blood, salt in water). This classification is still adopted, and comprises a broad number of wet materials with diverse compositions and physical structures. The authors emphasized that the selection of adequate drying techniques requires an individual approach for each material.

The versatility of drying over inert particles to produce different powders has been demonstrated along the latest decades. From bovine blood (Pham, 1983) to fruit pulps (Dantas et al., 2019; Medeiros et al., 2021) and fruit pomaces (Borges et al., 2016), a wide number of products have been successfully dried using inert particles in spout and fluidized beds. However, keeping wet spouted beds in continuous and stable operation is often a challenge and there are many issues associated to the limited control of the particle size distribution, the high-energy consumption required to keep spouting and the difficulty to scale-up (Freire, Ferreira, and Freire, 2011; Freire et al., 2012; Pallai, Szentmarjay, and Mujumdar, 2015).

Drying of a liquid film adhered to a solid surface comprehends the stages of coating, drying, cracking, film removal, and elutriation (Freire et al., 2012). During these processes, the paste undergoes complex changes that are difficult to predict and challenging to assess in real time. The magnitude of the adhesion forces at the interface between the coating and inert particles depends on many factors, including the paste composition, viscosity, surface tension, and other rheological properties (Freire, Ferreira, and Freire, 2011). A dryer operates stably whenever the rate at which the film is coated and removed is higher or equal to the rate of paste feeding. If these rates are unbalanced, paste accumulation will eventually trigger dynamic instabilities that ultimately will lead to a spouting or fluidization collapse. Even in wet beds with stable dynamics, controlling the product quality can be tricky due to the difficulty of performing reliable and accurate local measurements. Hence, in the last years, many researchers have focused on assessing the fluid dynamics on wet beds and on developing strategies to avoid instabilities either by altering the dryer configuration or by using additives to change the pastes' physical-chemical properties. There is also considerable interest in developing flexible and robust models aimed at improving the process control and enhancing the product quality.

With the advances in instrumentation and simulation techniques, considerable progress has been achieved in the analysis of fluid dynamics in wet beds. The influence of paste composition on the fluid dynamic behaviour and drying performance was consistently investigated and models were proposed to describe the fluid dynamics, heat, and mass transfer during drying with inert particles. A compilation of recent research and advances in this field is presented next.

It is agreed in the literature that the type of paste, solids concentration, and paste feed rate affects the fluid dynamic in wet beds. Depending on the kind of paste, operational conditions, and dryer configuration, dynamic parameters such as ΔP_{ms} and U_{ms} can be either smaller or larger than the ones found in dry beds operating under similar conditions (Patel et al., 1986; Schneider and Bridgwater, 1993; Passos and Mujumdar, 2000; Spitzner Neto, Cunha, and Freire, 2002; Almeida, Freire, and Freire, 2010).

All the others variables being the same, the maximum amount of water that the air can absorb is limited by the air temperature (Almeida et al., 2010). A dryer operates under steady conditions as long as the liquid phase is effectively removed to avoid paste or powder accumulation inside it. In such conditions, spouted and fluidized beds are well-mixed vessels with regard to heat and mass transfer.

According to Almeida et al. (2010), monitoring the dynamic behavior of a few key response variables in a SB dryer can be an effective strategy to identify the onset of dynamic instabilities. The authors observed a continuous decrease in ΔP_{ms} whenever the porosity of the annular region and the particle circulation rate were reduced due to air channelling. Under a moderate range of paste flowrate, the performance of the SB operating with water, sewage sludge, calcium carbonate suspensions (solid content from 3%–9%), and skim milk was similar regardless of the type of paste. This was attributed to the fact that the tested pastes were liquids with high water content and had low amount of fat or sugars in their compositions. The results of this study were the basis for further research on the development of a robust model aimed at predicting the dynamic behaviour of some key variables on drying, such as the outlet air temperature and relative humidity, and the powder moisture content. The proposed model was based on global energy and water mass balances, and an interphase coupling term was included in these balances to account for the simultaneous phenomena of heat exchange, water evaporation, and particle coating with the paste beds (Freire et al., 2012; Silva Costa et al., 2016). The coupling term was obtained from an ANN empirical model fitted for each paste. This hybrid lumped CSTR/ANN model predicted well the dynamic behaviour of the aforementioned variables during drying of sewage sludge, calcium carbonate suspensions, and skim milk. This flexible and simple to implement model can be easily extended for other types of pastes by including additional data into the ANN training procedure. A similar approach based on energy and mass balances was adopted by Dantas et al. (2018) to describe the dynamic behaviour of the outlet air temperature and humidity in SB drying of fruit pulp mixtures with intermittent feeding. The powder production rate, as well as the heat and mass transfer parameters, were estimated by empirically fitted equations and the model considered the accumulation of powder inside the dryer. The model fit well with the values of outlet air temperature and absolute humidity.

Intermittent feeding was recommended (Braga and Rocha, 2013; Dantas et al., 2018) to reduce the amount of water in the system and enhance stability during the drying process. An effective control requires models able to provide fast responses to different disturbances in the input variables. Moreira da Silva et al. (2019) proposed a semi-empirical lumped parameter model derived from overall mass and energy conservation balances to simulate paste drying of 5% concentrated calcium carbonate suspensions in a SB. The aim was to evaluate the model's responses to changes in the operating conditions and intermittent paste feeding. The outlet air temperature and humidity, as well as the powder moisture content predicted by the model were compared to experimental data obtained under a broad range of operating conditions. The authors confirmed that the intermittences in the paste feed rates contributed to enhancing stability and either avoided the spouting collapse or allowed operation for longer times as compared to continuous feeding under similar conditions. The model predicted well the temperature and air humidity but the powder moisture content was not predicted satisfactorily, even with a term to take into account the mass accumulation inside the dryer included in the mass balance.

Several factors contribute to make the prediction and control of output powder characteristics in drying with inert particles a huge challenge. Clustering and mass retention in the dryer vessel affect the production rate and powder characteristics. It is agreed that the operating conditions play a role in the powders' quality attributes such as solubility, texture, appearance, particle size, and particle size distribution (Kutsakova, 2004; Marreto, Freire, and Freitas, 2006; Benelli, Souza, and Oliveira, 2013). To assess the influence of operating conditions and mass retention on the powder's production and attributes during SB drying of calcium carbonate suspensions, Barros et al. (2019) analyzed the overall mass balance and evaluated the attributes of powders recovered from different parts of the dryer. Powder samples collected at the cyclone underflow, retained into the bed, and elutriated to the cyclone overflow, had their masses and PSD measured and quantified. The mass of solids collected at the cyclone underflow ranged from 11% to 53% of the total mass depending on the drying conditions. The authors found that the powders collected in different parts of the dryer were highly heterogeneous in sizes, with wide particle size distributions. Operating under a low paste feed rate and high temperature contributed to enhancing powder recovery at the cyclone underflow, reducing the amount of accumulated powder into the dryer, avoiding the onset of dynamic instabilities and favoured the production of a fine powder, with a mean size under 10 μm.

The control of powder moisture content is often an issue in spout/fluidized bed drying with inert particles because attrition and agglomeration during powder production cannot be easily assessed or controlled. The use of soft sensors, in which the powder moisture is inferred from an appropriate physical-mathematical model with real-time resolution and the measurement of other variables with high frequency and precision, is an alternative to address this problem. Vieira et al. (2019) incorporated the hybrid CSTR/ANN model into the data acquisition system of SBD used for the production of whole milk powder to obtain a soft sensor to predict the powder moisture content. The sensor was capable of estimating the milk powder moisture content when the dryer was submitted to disturbances on the air inlet temperature and paste inlet flow rate, but the model failed to describe paste accumulation within the bed. For this reason, the soft sensor overestimated the powder moisture content for long operation times.

Drying of tropical fruit pulps and related by-products in SBD with inert particles has received a great deal of attention in Brazil in recent years. Fruticulture is a major economic activity in the country but the reported losses in the production chain are up to 40% due to the fragile structure of most species combined to the tropical weather that favors fast deterioration of fresh crops (Medeiros et al., 2021). The intensive heat and mass transfer and low operating costs when compared to the spray dryers make the SB an appealing technique to process these products. However, the high moisture content and the reducing sugars in fruit composition make it difficult to keep a stable spouting during drying and may yield to products of poor quality product and low production rates. Local overheating and crusting are common issues to be faced in spouted bed drying of fruit pulps.

Braga and Rocha (2013) investigated spouted bed drying of a paste mixture of milk and blackberry (*Rubus sp*) pulp using drip-feeding or paste atomization over inert polystyrene particles. The best performance was observed by drying under 60°C with drip-feeding and maximum paste feed rate of 2 mL/min, a condition that yielded to powder moisture content lower than 3.5%, anthocyanin degradation less than 15% and powder production efficiency ranging from 44% to 64%. The authors reported significant powder retention inside the dryer under some conditions and noticed that the addition of milk to the pure blackberry pulp enhanced powder production and fluid dynamic stability.

The positive effect of adding dairy matrices, namely milk powder, reconstituted milk, and concentrated whey-protein as adjuvants in spouted bed drying of fruit pulps was also reported by Dantas et al. (2018), who investigated drying of acerola (*Malpighia emarginata* DC) pulp. The addition of dairy adjuvants contributed to reducing the equilibrium monolayer moisture of powdered samples and among the formulations tested, the one that best improved fluid dynamic and thermal performances was the mixture of acerola pulp and milk powder. Besides, the use of milk powder as adjuvant resulted in significant ascorbic acid retention (72.9%) after thermal processing.

In a review on spouted-bed drying of fruit pulps, Medeiros et al. (2021) addressed the use of adjuvants, intermittent feeding, and impact of the process on the phytochemicals of fruits. Besides the aforementioned fruits, this review lists other tropical and native fruits from Amazon and Cerrado biomes that have been successfully processed in SBs, such as bacaba (*Oenocarpus bacaba*), açaí (*Euterpe oleracea*), camu-camu (*Myrciaria dubia HBK McVaugh*), cubiu (*Solanum sessilifflorum* D), and graviola (*Annona muricata*). These fruits are all rich in bioactive compounds with pharmacological action, such as phenolic compounds, anthocyanins, and ascorbic acid. The production of fruit powders by drying the pulps in SBs has demonstrated to be an effective and low-cost alternative to reducing the losses in fruticulture and recovering bioactive constituents.

Spouted bed drying was also tested as a processing technique to recover bioactive constituents from tropical fruit pomaces (Borges et al., 2016). The residues were mixed with inert particles of polyethylene and the batches were dried during 40 min under air temperature of 70°C. The dried pomaces of acerola (*Malpighia glabra* L.), red pitanga (*Eugenia uniflora*), and jambolan (*Syzygium cumini*) presented attractive colors and moisture between 8.3 and 9.7 g/100 g. The authors found high concentrations of phenolic compounds and anthocyanins, and obtained relevant anti-oxidant activity in the water and ethanol extracts of the powders, showing that the pomaces preserved most of their nutritional and functional attributes.

6.3 DRYING PROCESS AND DRYER TYPES FOR PASTES

The complex rheological properties and the wide variety of paste-like materials (pastes, suspensions, solutions, slurries, and emulsions, all referred to henceforth in this chapter generically as pastes) prevent the recommendation of a "universal" type of dryer. Moreover, most organic and biological pastes are heat-sensitive materials, which define the temperature and residence time during the drying process as key aspects in the dryer selection. Therefore, the drying of pastes is a complex operation to handle in most industrial processes and the dryer selection is a challenging task. Kudra et al. (1989) have summarized some techniques used for the processing of paste-like materials, such as drying in a stationary layer (tray dryers), drying with material mixing, drying with a flowing film of a layer of material (band, film, and rotatory dryers), and dispersion drying (pneumatic, spray, or fluidized bed dryers). For the sake of clarity, the most common dryers for paste-like materials are described in this chapter based on the need or not to use inert material as a support medium in the drying process, as outlined in Figure 6.1.

Dryers without inert particles are more widely used in the industrial processing of pastes and solutions, either due to the industry resistance to technological innovations or due to product quality criteria. Thus, some of these dryers are described in the next section in which the industrial aspects are discussed with the recent developments and the perspectives for the drying of suspensions. Furthermore, many of these dryers

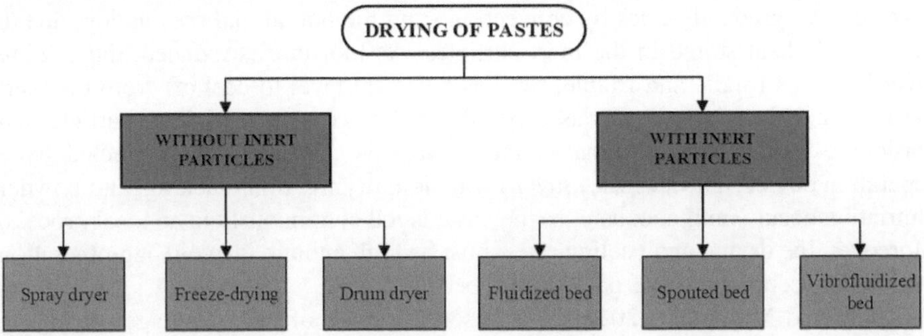

FIGURE 6.1 Some of the conventional and main dryers used in the drying of paste-like materials.

have been widely covered and reviewed in detail in recent literature (Filková, Huang, and Mujumdar, 2014; Tadeusz Kudra and Mujumdar, 2014; Liapis and Bruttini, 2014; Fu et al., 2020; Pinto et al., 2021). Therefore, the focus of this section is on conventional and new dryers with inert particles as a support medium in the drying process.

A variety of dryers are typically used to perform the drying of pastes on inert particles, of which classical fluidized bed, spouted bed, and vibrofluidized bed are the most popular ones (Figure 6.1). The mechanism behind the drying on inert particles is analogous, regardless of the dryer configuration, as illustrated in Figure 6.2. The paste is atomized or dropped into the bed by a nozzle or dropping device, coating the heated inert particles with a thin layer of moist material. The

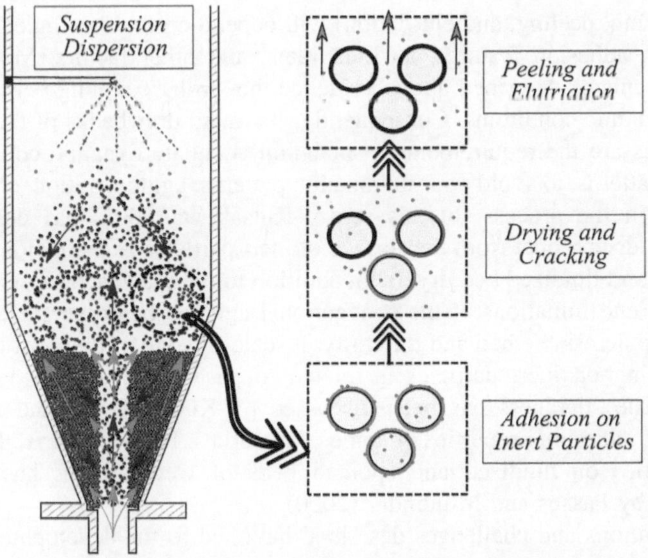

FIGURE 6.2 Drying mechanism of paste drying on inert particles in a conical spouted bed (adapted from Tellabide, 2020).

coating layer gradually dries by convection from the hot air and conduction, due to the sensible heat stored in the inert particles. As moisture is reduced, the coating layer becomes fragile and friable, reaches a critical level to peel off from the inert particle, and then is removed as a powder by successive particle-to-particle and particle-to-wall collisions. Finally, the powder is elutriated and collected by a separation device. The stages of layer formation, drying, film cracking, and powder elutriation occur simultaneously, as the paste is fed continuously in an ideal process. Moreover, the drying and peeling rates must be high enough to avoid agglomeration due to the accumulation of paste in the bed.

Kudra and Mujumdar (2014) have described some of the advantages of drying on inert particles compared to conventional and standard techniques, such as the spray dryer. The size of the inert particles, which may be from 20 to 40 times larger than that of the dispersed material dried, allows for higher gas velocity and, therefore, increased dryer productivity. The dispersion of material is improved due to the intensive motion of inert particles, which provides a fast and homogeneous dispersion, and thus coarse atomizers or dropping devices may be applied. Another advantage is related to the volumetric evaporation rate, as the use of a fluid bed of inert particles to dry slurries results in 15- to 17-fold higher rates than standard spray dryers under identical thermal conditions. Such advantages ensure even lower installation and operating costs than conventional methodologies without inert particles.

As presented in a previous review (Freire et al., 2012), the methodology based on inert particles has shown to be effective mainly for materials with high water content. However, this technique presents several variables and limitations that should be considered in the dryer selection for pastes processing, especially aimed at industrial applications. For instance, the time required for a complete cycle (coating, drying, peeling, and elutriation) will depend on the rheological properties of the paste, adhesion features, attrition rates, and other factors. Moreover, the drying mechanisms described above evidence the strong coupling between drying and fluid dynamic conditions. Consequently, the main drawbacks of the drying on inert particles are the requirements to maintain adequate dynamic conditions and circulation patterns, to avoid poor fluidization patterns, agglomeration, collapse, and instabilities in the process. In this sense, Kutsakova (2007) has described the removal of a dry product from the surface of inert particles as the major problem in spouted bed and fluidized bed dryers. In addition to the drawbacks described above and the inherent limitations of the conventional agitated bed dryers, such as high-pressure drop across the bed and difficulty in scale-up, one of the most challenging tasks in drying on inert particles is related to pastes with cohesive and sticky properties. The stickiness has been discussed by Kudra (2003) and reflects the cohesion and adhesion properties of some of materials. In this context, the effect of such properties on fluidized and spouted beds of wet particles has also been investigated by Passos and Mujumdar (2000).

The limitations and challenges described have led to the development of modified systems as an alternative to conventional agitated beds with inert particles (i.e., fluidized bed and spouted bed dryers). Some of these modifications for spouted bed dryers are represented in Figure 6.3.

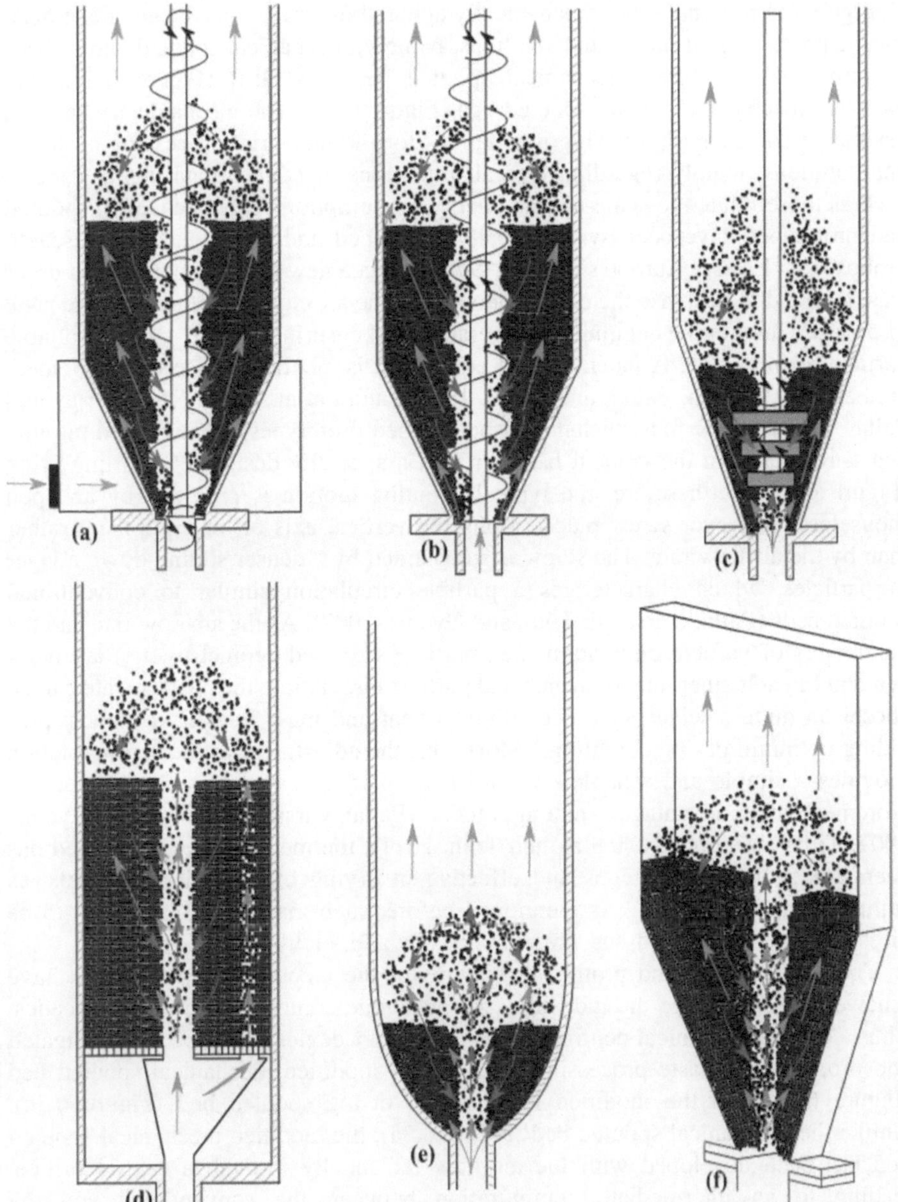

FIGURE 6.3 Some spouted bed configurations for drying of pastes on inert particles (adapted from Tellabide, 2020). (a) Conventional mechanical spouted bed. (b) Modified mechanical spouted bed. (c) Modified mechanical stirring spouted bed. (d) Dilute spin spouted bed. (e) Spouted bed with parabolic geometry. (f) 2-dimensional spouted bed.

Figure 6.3 presents some mechanically agitated dryers that have been developed for the processing of pastes and solutions, namely, mechanical spouted bed – MSB (Figure 6.3a), modified mechanical spouted bed – MMSB (Figure 6.3b), and modified mechanical stirring spouted bed (Figure 6.3c). The mechanically agitated dryers are considered promising alternatives for the industrial processing of pastes and solutions, mainly regarding typical limitations of conventional agitated beds, such as those related to scale-up and energy consumption. The mechanical spouted bed prototypes have been systematically developed and investigated by research groups from Eastern Europe since the 1970s, when a new type of spouted bed dryer was designed to decrease the energy demand of the air blower, eliminating the peak of pressure drop at the beginning of operating and ensuring intense and controllable particle recirculation. A patent of this equipment is mentioned elsewhere (Németh et al., 1983; Németh, Pallai, and Aradi, 1983; Kudra et al., 1989; Szentmarjay and Pallai, 1989). In the first mechanical spouted bed prototypes, the air is fed into the bed tangentially in the conical base through a specially designed "whirling" ring (Figure 6.3a). Furthermore, the typical spouting motion is provided by an open (houseless) conveyor screw placed along the vertical axis of the equipment rather than by the air flow rate. The screw is surrounded by a denser sliding down a layer of particles, which characterizes a particle circulation similar to conventional spouted beds (Pallai-Varsányi, Tóth, and Gyenis, 2007). As the air-flow rate and the parameters of particle circulation (i.e., particle size, bed geometry, etc.) are independent of each other due to mechanical particle circulation, the air-flow rate can be chosen in quite a wide range according to heat and mass transfer demands, providing optimum drying conditions. Moreover, the adjustment of the screw rotation provides a simple and reliable control of the particle circulation rate. Given the aforementioned advantages, among others (Pallai-Varsányi, Tóth, and Gyenis, 2007; Tibor Szentmarjay, Pallai, and Tóth, 2011), the mechanical spouted bed has been especially advantageous and effective in drying of paste-like materials, as summarized in Table 6.1. As mentioned before, such materials are difficult to be directly spouted/fluidized and present cohesive and sticky properties.

These advantages and promising features of the mechanical spouted bed have renewed the interest of the industrial and academic community in recent decades. Thus, several mechanical configurations and stirrer devices have been investigated and proposed for paste processing, such as the modified mechanical spouted bed (Figure 6.3b) and the modified mechanical stirring spouted bed (Figure 6.3c). Unlike the mechanical spouted bed (Figure 6.3a), the modified mechanical spouted bed has been developed with the air flow fed axially through a central orifice, leading to an intermediate configuration between the conventional and the mechanical spouted bed dryers. This modification results in an experimental unit suitable to operate as either a mechanically or a conventional spouted bed, providing a feasible and flexible configuration with a wide operating range for practical applications in the processing of pastes and particulate solids (Brito et al., 2018; Sousa et al., 2019). When handling pastes of complex composition with high viscosity or when drying generates viscous-elastic films, such as those with fruit pulp (Reyes, Díaz, and Blasco, 1998), the use of different stirrer devices is an interesting alternative to the conveyor screw. In this sense, Reyes and Vidal (2000) and Barros et al. (2020)

TABLE 6.1
Summary of Some Applications Involving Mechanical Spouted Beds for Drying of Pasty Materials

	Inert Particle (Size)	Pasty Material
Szentmarjay and Pallai (1989)	Glass spheres (6.0 mm)	Cobalt carbonate Zinc carbonate
Szentmarjay et al. (1994)	Ceramic spheres (6.6 mm and 7.4 mm)	Cobalt carbonate
Szentmarjay et al. (1995)	Hostaform spheres (7.6 mm)	Brewery yeast suspension
Szentmarjay et al. (1996)	Hostaform spheres (7.6 mm)	Brewery yeast suspension starch suspension
Szentmarjay et al. (1996)	Glass spheres (4.0 mm) Hostaform spheres (7.6 mm)	Tomato concentrate Brewery yeast suspension
Szentmarjay and Pallai (2000)	Teflon particles (12.0 mm)	AlO (OH) suspension
Pallai et al. (2001)	Hostaform spheres (7.6 mm)	Potato pulp
Pallai-Varsányi et al. (2007)	Hostaform spheres (7.6 mm)	AlO (OH) suspension Tomato concentrate

have proposed the mechanically stirred spouted bed and the modified mechanical stirring spouted bed (Figure 6.3c), respectively. Both dryers are designed with different stirrer devices that can provide bed agitation and avoid the formation of agglomerates. Thus, the air flow required to reach a suitable dynamic condition is significantly decreased due to these stirrer devices, which means less energy demand and operating costs.

In the processing of pastes, such internal stirrers are also widely used in agitated beds other than the spouted bed, such as fluidized bed dryers. Interesting stirrer devices are investigated by Bait et al. (2011), who have employed three different agitators in fluidized beds for drying of cohesive particles at low air velocity, namely, straight-blade, pitch-blade, and helical ribbon-type agitators. These authors have obtained benefits related to scale-up and homogeneous mixing provided by the agitators, which are important aspects to be considered in any practical industrial application. Reyes et al. (2001) have also used stirrer devices in fluidized bed and spouted bed dryers by the implementation of a blade agitator and a paddle agitator, respectively, in the base section connected through a vertical axis. Overall, these authors have observed that the incorporation of mechanical agitation in the drying of liquid substrates in fluidized or spouted beds of inert particles increased significantly their drying capacity. Kudra and Mujumdar (2009) have described in detail several other interesting drying technologies based on inert particles for pasty-like materials, which evidences the variety of dryers for these materials and the numerous opportunities for innovation, especially regarding the complexity and the particularities of each pasty-like material. Thus, the recent developments, the perspectives, and industry-academia integration concerning the drying of solutions and pastes are discussed in the following section.

6.4 INDUSTRIAL AND ACADEMIC PERSPECTIVES

Drying of suspensions, solutions, sludges, and/or paste-like materials consists of a common way of producing powder and granules, with a vital role in several industries sectors, such as chemical, pharmaceutical, and food industries. Most of these industries involve the processing of heat-sensitive and high value-added materials, which demand high-quality dry products. Furthermore, drying is a highly energy-intensive process and dryers are considered one of the most energy-consuming industrial equipment, with a significant impact on costs and environmental issues. However, despite the numerous innovative drying technologies proposed by academia over the past decades, industrial innovations remain challenging and conventional technologies still dominate the market. Thus, classical dryers are usually adopted for the industrial processing of pastes and solutions, such as spray, drum, and freeze dryers.

In the food industry, spray drying is usually employed on an industrial scale for the processing of milk, whey, egg, soy protein, coffee powder, starch powder, tomato purees, among others (Filková, Huang, and Mujumdar, 2014; Fu et al., 2020). A typical spray-drying process consists of liquid atomization into a hot air flow in a drying chamber, providing the drying of droplets with the formation of solid particle agglomerates, which are subsequently separated from the stream by a variety of methods (Filková, Huang, and Mujumdar, 2014; Walters et al., 2014). A spray dryer provides a dry product of controllable particle specific properties (size, shape, form, moisture, etc.), which is an important advantage for the food industry due to the drying of literally hundreds of variants of pastes and solutions (Filková, Huang, and Mujumdar, 2014; Fu et al., 2020). Furthermore, spray dryers present other advantages related to scale-up design and flexible continuous operation, which are considered key aspects for industrial applications. A detailed description of the spray-drying technology is beyond the scope of this chapter and can be found in Fu et al. (2020) and Filková, Huang, and Mujumdar (2014).

In the pharmaceutical industry, freeze drying (or lyophilization) is the most widely utilized technology for transforming a wide range of aqueous and non-aqueous solutions of bioactive substances (antibiotics, bacteria, vaccines, chemicals, biotechnological products, etc.) into a solid and stable state, as a dry powder (Devahastin and Jinorose, 2020). The principle of freeze drying is based on the remotion of a solvent (normally water) from a frozen solution by sublimation and desorption (Liapis and Bruttini, 2014; Pinto et al., 2021). Despite freeze drying providing the highest quality product obtained by any drying method, with several commercially approved products, spray drying is considered the most mature alternative technology for industrial implementation in the production of many pharmaceutical products, such as those described by Thorat, Sett, and Mujumdar (2020) and Pinto et al. (2021).

Although conventional dryers still dominate the market, such technologies have several drawbacks and limitations. Freeze drying, for example, is characterized by difficulties in the processing of large quantities of material and in controlling the particle properties (Pinto et al., 2021), whereas spray drying usually has low thermal efficiency and product deposit on the drying chamber, leading to degraded products

(Filková, Huang, and Mujumdar, 2014). In addition, both technologies require relatively high installation and operating costs compared to other drying technologies. This limitation makes the industrial application unfeasible for the processing of pastes and solutions with low value-added and negligible quality demands, such as sludges, waste products, and commodity chemicals. These aspects, coupled with the wide variety of dryers and manufactured products, provide ample opportunities for innovation and have constantly motivated academia to develop new and hybrid technologies for the processing of pastes and solutions. Drying on inert particles, pulse-combustion dryers, superheated steam dryers, intermittent drying, among others, are some of these technologies, which are described in detail by Kudra and Mujumdar (2014).

As discussed throughout this chapter, drying on inert particles is a relatively novel technology for the processing of pastes, slurries, and solutions. This technology is recognized as an efficient method to obtain powdery products, based on drying of pasty on the surface of heated inert particle, which serves as a carrier for the liquid film and a heat transfer medium. Originally proposed in the late 1970s at the Leningrad Institute of Technology as an alternative to eliminate the constraints of spray, drum, and paddle dryers (Mathur and Epstein, 1974; Arsenijevic, Grbavcic, and Garic-Grulovic, 2002), the drying of solutions in beds of inert particles has been applied in several classical dryers, namely, spouted bed, fluidized bed, and vibrated fluidized bed. As described in the previous section, many configurations and modifications of these dryers have been developed by academia to provide feasible industrial applications for a wide variety of paste-like materials. A partial list of these applications is also presented by Freire, Ferreira, and Freire (2011).

Typical pulse combustion dryers consist basically of a drying chamber and a pulse combustor (PC) that provides an intermittent (pulse) combustion of the solid, liquid, or gaseous fuel, unlike the continuous combustion in conventional burners. Besides the heat supply, the pulse combustor generates large-amplitude high-frequency pressure pulsations within a drying chamber (spray dryer, rotatory dryer, pneumatic dryer, spouted bed dyer, among others), thus improving the drying rate (Kudra, 2008). Meng, De Jong, and Kudra (2016) have also described many other advantages of the P-C dryer that characterize such technology as one of the most promising drying techniques compared to classical dryers. The application of the P-C dryer for the processing of slurries and suspensions has been attempted only in the last decades, intensifying industrial applications mainly in spray drying, fluid bed drying, and flash drying (Zhonghua and Mujumdar, 2006; Zhonghua et al., 2012). Although the P-C dryer is not yet widely applied, this technology has been considered commercially viable based on information provided by the three major dryer manufacturers, namely, Pulse Drying Systems (Portland, OR), Novadyne Ltd. (Hastings, ON, Canada), and Pulse Combustion Systems™ (San Rafael, CA). A detailed description of some of these commercial P-C dryers can be found in Meng, De Jong, and Kudra (2016) and Kudra (2008).

As briefly presented above, there are numerous opportunities for innovation in the drying of pastes and solutions, providing a promising perspective of interaction between academia and industry. However, the adoption of new ideas from academia in industrial practice remains at a low level. Mujumdar (2011, 2013b, 2013a, 2018a, 2018b) has constantly addressed some of the issues involved in academia-industry collaboration in drying research. The main issue is related to the distinct motivations, i.e., the knowledge generation and dissemination and the educating researchers are the goals of academia, whereas profit-making is the main goal of the industry. Thus, the time scale for academic research and industry are significantly different, with the latter rightfully interested in faster turnaround. Consequently, there is usually strong resistance from the industry to fund academic projects. Furthermore, the need for academia to publish scientific results is often a limitation in industry-academia collaboration due to issues related to intellectual property rights. Nevertheless, academia-industry interaction has remarkable potential to intensify innovation, mainly in the processing of pastes and solutions, providing new designs and configurations, better control of dryers, reduction of carbon footprint, etc. Therefore, efforts must be made to promote academia-industry interaction in a mutually beneficial manner.

REFERENCES

Almeida, A. R. F., F. B. Freire, and J. T. Freire. 2010. "Transient Analysis of Pasty Material Drying in a Spouted Bed of Inert Particles." *Drying Technology* 28 (3): 330–340. 10.1080/07373931003627189

Arsenijevic, Z., Z. Grbavcic, and R. Garic-Grulovic. 2002. "Drying of Solutions and Suspensions in the Modified Spouted Bed with Draft Tube." *Thermal Science* 6 (2): 47–70. 10.2298/tsci0202047a

Bait, R. G., S. B. Pawar, A. N. Banerjee, A. S. Mujumdar, and B. N. Thorat. 2011. "Mechanically Agitated Fluidized Bed Drying of Cohesive Particles at Low Air Velocity." *Drying Technology* 29 (7): 808–818. 10.1080/07373937.2010.541574

Barros, J. P. A. A., M. C. Ferreira, and J. T. Freire. 2019. "Spouted Bed Drying on Inert Particles: Evaluation of Particle Size Distribution of Recovered, Accumulated and Elutriated Powders." *Drying Technology*: 1–12. 10.1080/07373937.2019.1656644

Barros, João Pedro Alves de Azevedo, R. C. de Brito, F. B. Freire, and J. T. Freire. 2020. "Fluid Dynamic Analysis of a Modified Mechanical Stirring Spouted Bed: Effect of Particle Properties and Stirring Rotation." *Industrial & Engineering Chemistry Research* 59 (37): 16396–16406. 10.1021/acs.iecr.0c03139

Benelli, L., C. R. F. Souza, and W. P. Oliveira. 2013. "Spouted Bed Performance on Drying of an Aromatic Plant Extract." *Powder Technology* 239: 59–71. 10.1016/j.powtec.2013.01.058

Borges, K. C., J. C. Azevedo, Maria De Fátima Medeiros, and R. T. P. Correia. 2016. "Physicochemical Characterization and Bioactive Value of Tropical Berry Pomaces after Spouted Bed Drying." *Journal of Food Quality* 39 (3): 192–200. 10.1111/jfq.12178

Braga, M. B., and S. C. S. Rocha. 2013. "Drying of Milk-Blackberry Pulp Mixture in Spouted Bed." *Canadian Journal of Chemical Engineering* 91 (11): 1786–1792. 10.1002/cjce.21918

Brito, R. C., R. C. Sousa, R. Béttega, F. B. Freire, and J. T. Freire. 2018. "Analysis of the Energy Performance of a Modified Mechanically Spouted Bed Applied in the Drying of Alumina and Skimmed Milk." *Chemical Engineering and Processing – Process Intensification* 130. 10.1016/j.cep.2018.05.014

Dantas, Suziani Cristina de Medeiros, Severino Mosinho de Pontes Júnior, Fábio Gonçalves Macêdo de Medeiros, Luiz Carlos Santos Júnior, Odelsia Leonor Sanchez de Alsina, and Maria de Fátima Dantas de Medeiros. 2019. "Spouted-bed Drying of Acerola Pulp (Malpighia Emarginata DC): Effects of Adding Milk and Milk Protein on Process Performance and Characterization of Dried Fruit Powders." *Journal of Food Process Engineering*, 42(6): 1–13. 10.1111/jfpe.13205

Dantas, Thayse Naianne Pires, Francisco Canindé Moraes Filho, Josilma Silva Souza, Jackson Araújo de Oliveira, Sandra Cristina dos Santos Rocha, and Maria de Fátima Dantas de Medeiros. 2018. "Study of Model Application for Drying of Pulp Fruit in Spouted Bed with Intermittent Feeding and Accumulation." *Drying Technology* 36 (11): 1349–1366. 10.1080/07373937.2017.1402785

Devahastin, S., and M. Jinorose. 2020. "A Concise History of Drying." In *Drying Technologies for Biotechnology and Pharmaceutical Applications*, edited by S. Ohtake, K.-I. Izutsu, and D. Lechuga-Ballesteros, 9–21. Weinheim, Germany: Wiley-VCH.

Filková, I., L. X. Huang, and A. S. Mujumdar. 2014. "Industrial Spray Drying Systems." In *Handbook of Industrial Drying*, edited by A. S. Mujumdar, Fourth Ed., 191–226. New York: CRC Press Taylor & Francis.

Freire, J. T., M. C. Ferreira, and F. B. Freire. 2011. "Drying of Solutions, Slurries, and Pastes." In *Spouted and Spout-Fluid Beds: Fundamentals and Applications*, edited by N. Epstein and J. R. Grace, First Ed., 206–221. Cambridge: Cambridge University Press.

Freire, J. T., M. C. Ferreira, F. B. Freire, and B. S. Nascimento. 2012. "A Review on Paste Drying with Inert Particles as Support Medium." *Drying Technology* 30 (4): 330–341. 10.1080/07373937.2011.638149

Fu, N., J. Xiao, M. W. Woo, and X. D. Chen. 2020. *Frontiers in Spray Drying*. *Frontiers in Spray Drying*. Florida: CRC Press. 10.1201/9780429429859

Kudra, T. 2003. "Sticky Region in Drying – Definition and Identification." *Drying Technology* 21 (8): 1457–1469. 10.1081/DRT-120024678

Kudra, T. 2008. "Pulse-Combustion Drying: Status and Potentials." *Drying Technology* 26 (12): 1409–1420. 10.1080/07373930802458812

Kudra, T., and A. S. Mujumdar. 2009. *Advanced Drying Technologies*. 2nd Ed. Boca Raton: CRC Press Taylor & Francis.

Kudra, T., and A. S. Mujumdar. 2014. "Special Drying Techniques and Novel Dryers." In *Handbook of Industrial Drying*, edited by A. S. Mujumdar, Fourth Ed., 433–489. New York: CRC Press Taylor & Francis.

Kudra, T., E. Pallai, Z. Bartczaki, and M. Peter. 1989. "Drying of Paste-like Materials in Screw-Type Spouted-Bed and Spin-Flash Dryers." *Drying Technology* 7 (3): 583–597. 10.1080/07373938908916612

Kutsakova, V. E. 2004. "Drying of Liquid and Pasty Products in a Modified Spouted Bed of Inert Particles." *Drying Technology* 22 (10): 2343–2350. 10.1081/DRT-200040018

Kutsakova, V. E. 2007. "Effect of Inert Particles Properties on Performance of Spouted Bed Dryers." *Drying Technology* 25: 617–620. 10.1080/07373930701250062

Liapis, A. I., and R. Bruttini. 2014. "Freeze Drying." In *Handbook of Industrial Drying*, edited by A. S. Mujumdar, Fourth Ed., 259–282. Florida: CRC Press Taylor & Francis.

Marreto, R. N., J. T. Freire, and L. A. P. Freitas. 2006. "Drying of Pharmaceuticals: The Applicability of Spouted Beds." *Drying Technology* 24 (3): 327–338. 10.1080/07373930600564324

Mathur, K. B., and N. Epstein. 1974. *Spouted Beds*. New York: Academic Press.

Medeiros, F. G. M., I. P. Machado, T. N. P. Dantas, S. C. M. Dantas, O. L. S. Alsina, and M. F. D. Medeiros. 2021. "Spouted Bed Drying of Fruit Pulps: A Case Study on Drying of Graviola (Annona Muricata) Pulp." In *Transport Processes and Separation Technologies*, edited by J. M. P. Q. Delgado and A. G. Barbosa de Lima, 105–150. Switzerland: Springer.

Meng, X., W. De Jong, and T. Kudra. 2016. "A State-of-the-Art Review of Pulse Combustion: Principles, Modeling, Applications and R&D Issues." *Renewable and Sustainable Energy Reviews* 55: 73–114. 10.1016/j.rser.2015.10.110

Moreira da Silva, C. A., Maria do Carmo Ferreira, F. B. Freire, and J. T. Freire. 2019. "Analysis of the Dynamics of Paste Drying in a Spouted Bed." *Drying Technology* 37 (7): 876–884. 10.1080/07373937.2018.1471699

Mujumdar, A. S. 2011. "Editorial: Industrial Innovation—Is Academic Research a Significant Influence?" *Drying Technology* 29 (6): 609–609. 10.1080/07373937.2011.560794

Mujumdar, A. S. 2013a. "Editorial: On Industry–Academia Interaction: Hallmark of Engineering Research." *Drying Technology* 31 (9): 965–965. 10.1080/07373937.2013.812472

Mujumdar, A. S. 2013b. "Editorial: How Industry May Benefit from Interaction with Academia." *Drying Technology* 31 (10): 1083–1083. 10.1080/07373937.2013.812766

Mujumdar, A. S. 2018a. "Editorial: On Academia–Industry Collaboration in Drying Research." *Drying Technology* 36 (7): 763. 10.1080/07373937.2017.1350626

Mujumdar, A. S. 2018b. "Role of Academia in Industrial Developments." *Drying Technology*: 1–1. 10.1080/07373937.2018.1453442

Németh, J., E. Pallai, and E. Aradi. 1983. "Scale-up Examination of Spouted Bed Dryers." *The Canadian Journal of Chemical Engineering* 61 (3): 419–425. 10.1002/cjce.5450610324

Németh, J., E. Pallai, M. Péter, and R. Törös. 1983. "Heat Transfer in a Novel Type Spouted Bed." *The Canadian Journal of Chemical Engineering* 61 (3): 406–410. 10.1002/cjce.5450610322

Pallai, E., T. Szentmarjay, and A. S. Mujumdar. 2015. "Spouted Bed Drying." In *Handbook of Industrial Drying*, edited by A. S. Mujumdar, Fourth Ed., 351–370. Boca Raton: CRC Press.

Pallai, E., T. Szentmarjay, and E. Szijjártó. 2001. "Effect of Partial Processes of Drying on Inert Particles on Product Quality." *Drying Technology* 19 (8): 2019–2032. 10.1081/DRT-100107286

Pallai-Varsányi, E., J. Tóth, and J. Gyenis. 2007. "Drying of Suspensions and Solutions on Inert Particle Surface in Mechanically Spouted Bed Dryer." *China Particuology* 5 (5): 337–344. 10.1016/j.cpart.2007.06.003

Passos, M. L., and A. S. Mujumdar. 2000. "Effect of Cohesive Forces on Fluidized and Spouted Beds of Wet Particles." *Powder Technology* 110 (3): 222–238. 10.1016/S0032-5910(99)00278-8

Patel, K., J. Bridgwater, C. G. J. Baker, and T. Schneider. 1986. "Spouting Behavior of Wet Solids." In *Drying' 86*, edited by A. S. Mujumdar. New York: Hemisphere Publishing.

Pham, Q. T. 1983. "Behaviour of a Conical Spouted-bed Dryer for Animal Blood." *The Canadian Journal of Chemical Engineering* 61 (3): 426–434. 10.1002/cjce.5450610325

Pinto, J. T., E. Faulhammer, J. Dieplinger, M. Dekner, C. Makert, M. Nieder, and A. Paudel. 2021. "Progress in Spray-Drying of Protein Pharmaceuticals: Literature Analysis of Trends in Formulation and Process Attributes." *Drying Technology*: 1–32. 10.1080/07373937.2021.1903032

Reyes, A., G. Díaz, and R. Blasco. 1998. "Slurry Drying in Gas-Particle Contactors: Fluid-Dynamics and Capacity Analysis." *Drying Technology* 16 (1–2): 217–233. 10.1080/07373939808917400

Reyes, A., G. Díaz, and F. H. Marquardt. 2001. "Analysis of Mechanically Agitated Fluid-Particle Contact Dryers." *Drying Technology* 19 (9): 2235–2259.

Reyes, A., and I. Vidal. 2000. "Experimental Analysis of a Mechanically Stirred Spouted Bed Dryer." *Drying Technology* 18 (1–2): 341–359.

Schneider, T., and J. Bridgwater. 1993. "THE Stability of Wet Spouted Beds." *Drying Technology* 11 (2): 277–301. 10.1080/07373939308916820

Silva Costa, A. B., F. Bentes Freire, J. T. Freire, and M. C. Ferreira. 2016. "Modelling Drying Pastes in Vibrofluidized Bed with Inert Particles." *Chemical Engineering and Processing: Process Intensification* 103 (May): 1–11. 10.1016/j.cep.2015.09.012

Sousa, R. C., M. C. Ferreira, H. Altzibar, F. B. Freire, and J. T. Freire. 2019. "Drying of Pasty and Granular Materials in Mechanically and Conventional Spouted Beds." *Particuology* 42 (February): 176–183. 10.1016/j.partic.2018.01.006

Spitzner Neto, P. I., F. O. Cunha, and J. T. Freire. 2002. "Effect of the Presence of Paste in a Conical Spouted Bed Dryer With Continuous Feeding." *Drying Technology* 20 (4–5): 789–811. 10.1081/DRT-120003758

Strumiłło, C., A. Markowski, and W. Kaminski. 1983. "Modern Developments in Drying of Pastelike Materials." In *Advances in Drying Vol. 2*, edited by A. S. Mujumdar, 193–231. Washington, DC: McGraw-Hill.

Szentmarjay, T., and E. Pallai. 2000. "Drying Experiments with ALO(OH) Suspension of High Purity and Fine Particulate Size to Design an Industrial Scale Dryer." *Drying Technology* 18 (3): 759–776. 10.1080/07373930008917736

Szentmarjay, T., and E. Pallai. 1989. "Drying of Suspensions in a Modified Spouted Bed Drier with an Inert Packing." *Drying Technology* 7 (3): 523–536. 10.1080/07373938908916607

Szentmarjay, T., E. Pallai, and Z. Regényi. 1996. "Short-Time Drying of Heat-Sensitive, Biologically Active Pulps and Pastes." *Drying Technology* 14 (9): 2091–2115. 10.1080/07373939608917197

Szentmarjay, T., E. Pallai, and A. Szalay. 1995. "Drying Process on Inert Particles in Mechanically Spouted Bed Dryer." *Drying Technology* 13 (5–7): 1203–1219. 10.1080/07373939508917017

Szentmarjay, T., E. Pallai, and J. Tóth. 2011. "Mechanical Spouting." In *Spouted and Spout-Fluid Beds: Fundamentals and Applications*, 297–304.

Szentmarjay, T., A. Szalay, and E. Pallai. 1994. "Scale-up Aspects of the Mechanically Spouted Bed Dryer with Inert Particles." *Drying Technology* 12 (1–2): 341–350. 10.1080/07373939408959960

Tellabide, M. 2020. "Partikula Finen Tratamendurako Iturri Bilgailudun Ohantze Konikoen Hidrodinamika." University of the Basque Country.

Thorat, B. N., A. Sett, and A. S. Mujumdar. 2020. "Drying of Vaccines and Biomolecules." *Drying Technology*: 1–23. 10.1080/07373937.2020.1825293

Vieira, G. N. A., M. Olazar, J. T. Freire, and F. B. Freire. 2019. "Real-Time Monitoring of Milk Powder Moisture Content during Drying in a Spouted Bed Dryer Using a Hybrid Neural Soft Sensor." *Drying Technology* 37 (9): 1184–1190. 10.1080/07373937.2018.1492614

Walters, R. H., B. Bhatnagar, S. Tchessalov, K. I. Izutsu, K. Tsumoto, and S. Ohtake. 2014. "Next Generation Drying Technologies for Pharmaceutical Applications." *Journal of Pharmaceutical Sciences* 103 (9): 2673–2695. 10.1002/jps.23998

Zhonghua, W., W. Long, L. Zhanyong, and A. S. Mujumdar. 2012. "Atomization and Drying Characteristics of Sewage Sludge inside a Helmholtz Pulse Combustor." *Drying Technology* 30 (10): 1105–1112. 10.1080/07373937.2012.683122

Zhonghua, W., and A. S. Mujumdar. 2006. "R&D Needs and Opportunities in Pulse Combustion and Pulse Combustion Drying." *Drying Technology* 24 (11): 1521–1523. 10.1080/07373930600961520

7 Rotary Dryers

Fluid Dynamics Aspects and Modelling

Marcos Antonio de Souza Barrozo
School of Chemical Engineering, Federal University of Uberlândia, Uberlândia, Brazil

Dyrney Araujo dos Santos
Institute of Chemistry, Federal University of Goiás, Goiânia, Brazil

Cláudio Roberto Duarte
School of Chemical Engineering, Federal University of Uberlândia, Uberlândia, Brazil

Sullen Mendonça Nascimento
Department of Engineering, Federal University of Lavras, Lavras, Brazil

CONTENTS

DOI: 10.1201/9781003207108-7

7.1 INTRODUCTION

Rotary dryers have been used in many industrial sectors, including the cement, fertilizer, and minerals industries (Lisboa et al., 2007), because of their high processing capacity and their greater flexibility in handling a wide variety of materials than other types of dryers (Felipe et al., 2003). If correctly designed and operated, they provide high thermal efficiencies.

There are two main types of configurations of a rotary drum: with flights and without flights. For both configurations, the fluid dynamics aspects strongly influence the drying performance. In this chapter, we present some important analysis about the fluid dynamics behavior in rotary drums with and without flights.

7.2 FLIGHTED ROTARY DRUM

7.2.1 Introduction

Flighted rotary dryer is a type of dryer that is commonly used in industry to dry particulate solids (Keey, 1972). It is made of a long cylindrical shell that is rotated. The shell is usually slightly inclined to the horizontal to induce solids flow from one end of the dryer to the other, as can be observed in Figure 7.1. In direct heated rotary dryers, a hot gas flowing through the equipment provides the heat required for vaporization of the water.

To promote gas-solid contact, most direct rotary dryers have flights (see Figure 7.1), placed parallel along the length of the shell. As the dryer rotates, the solids are picked up by the flights and are conveyed for a certain distance around the periphery before dislodging and falling back as a raining curtain through a hot air stream. Thus, the material inside the dryer is classified into three different phases: the dense phase at the bottom of the cylinder; the flight phase (passive), where the materials are lifted by the flights; and the airborne phase (active), where the material is directly exposed to the airstream. In conventional cascading rotary dryers, the particulate material mostly dries during the falling period (active phase). Therefore, good flight design is essential to promote the gas-solid contact that is required for rapid and homogeneous drying (Revol et al., 2001). Because of this, rotary dryers represent one of the greatest

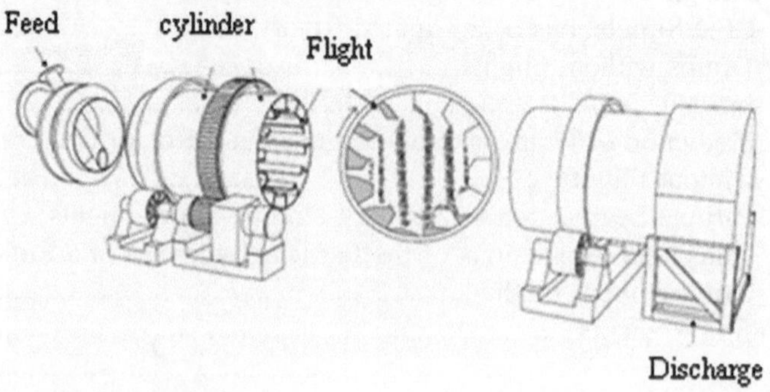

FIGURE 7.1 The rotary dryer.

challenges in theoretical modeling of dryers (Britton et al., 2006). The complex combination of particles being lifted by the flights, sliding and rolling, then falling in spreading cascades through an airstream and re-entering the bed at the bottom, possibly with bouncing and rolling, is very difficult to analyze (Kemp, 2004).

The performance of a rotary dryer critically depends on the drum loading, which is the total amount of solid material fed into the drum (Santos et al., 2015). The feed rate influences the amount of solids in the flight phase (passive) and in the airborne phase (active), which affects the residence time and the heat and mass transfers (Santos et al., 2015).

Despite its simplicity and flexibility, conventional rotary dryer units represent a significant cost to industry. The scale-up of rotary dryers (Lisboa et al., 2007) from pilot scale and the control of product quality during processing disturbances remain complex and poorly described. Thus, because of the complexity in estimating some parameters, most rotary dryers are overdesigned and as a consequence the final product can be over-dried and over-heated, wasting thermal energy. In this chapter, we discuss some important aspects of fluid dynamics in rotary drums, which directly influence the drying performance of these devices.

7.2.2 Geometric Equations

A dryer may incorporate one or more different types of flights, in sufficient number, that must be distributed across the drum (Silverio et al., 2015). The number and format of flights influence the amount of material present in the rotary dryer. Perry and Green (1999) suggest that the volume occupied by the load of solids in the rotary dryer should be between 10% and 15% of the total dryer volume.

To illustrate a way to obtain equations to predict the holdup based on geometric characteristics of the flights, in this chapter we consider a two-segment flight (Lisboa et al., 2007). Figure 7.2 presents a scheme of the flights with two segments, indicating the main dimensions and variables.

The flight presented in Figure 7.2, can be characterized by the lengths of segment 1 (l') and segment 2 (l), the angle between the two segments (α_A) and the circle radius

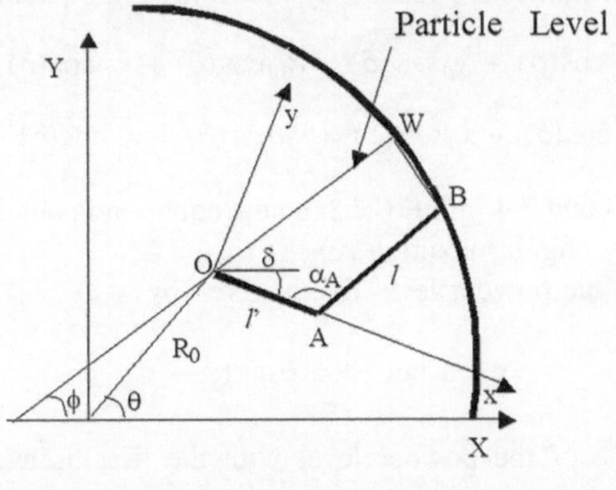

FIGURE 7.2 Scheme of a two-segment flight.

(R_o) formed by the line between the edge of the flight (O) and the center of the rotary drum. Two sets of Cartesian coordinates are considered. The origin (x,y) of the set is at the flight lip, the x-axis is located along the first segment and the y-axis, perpendicular to the x-axis. This set of coordinates moves as the flight rotates. The origin of the stationary (X,Y) set is on the drum axis with the x-axis being horizontal (Figure 7.2).

Schofield and Glikin (1962) linked the dynamic angle of repose, ϕ (the angle formed by the level of material in the flights and the horizontal line) to the dynamic friction angle of the material (μ), the angular position of the flight edge (θ), the radial position of the flight edge (R_o), and the drum rotation speed (ω), using the following equation:

$$\tan\phi = \frac{\mu + R_0\frac{\omega^2}{g}(\cos\theta - \mu sen\theta)}{1 - R_0\frac{\omega^2}{g}(sen\theta - \mu\cos\theta)} \tag{7.1}$$

In order to calculate the area occupied by the material in the flights, the coordinates of points A and B are first calculated (see Figure 7.2), the angle δ between the two sets of coordinates is evaluated, and the volume of material is finally obtained.

Equations for the two-segment flights may be obtained this way:

Segment coordinates: 1 ($y_1 = 0$) and segment 2 ($y_2 = a_2 + b_2x$); with: $a_2 = x_A \text{tang}(\alpha_A)$ and $b_2 = -\text{tang}(\alpha_A)$.

The coordinates of points A and B are given by

- Point A: $x_A = l'$ and $y_A = 0$.
- Point B: $x_B = x_A - l\cos(\alpha_A)$ and $y_B = l\text{sen}(\alpha_A)$.

In the stationary set of coordinates, the position of point B must satisfy the following equation since it is located on the wall of the drum of radius R:

$$X_B^2 + Y_B^2 = R^2 \tag{7.2}$$

The two sets of coordinates are related by the following equations:

$$X_B = X_0 + x_B\cos(\delta) - y_B sen(\delta) = R_0\cos(\theta) + x_B\cos(\delta) - y_B sen(\delta) \tag{7.3}$$

$$Y_B = Y_0 + y_B\cos(\delta) - x_B sen(\delta) = R_0 sen(\theta) + y_B\cos(\delta) - x_B sen(\delta) \tag{7.4}$$

Substituting Eqs. 7.3 and 7.4 into Eq. 7.2, a new equation is obtained, which can be solved for δ, for any angular position (θ).

The equation for the powder level line is given by

$$y = x\tan(\gamma) = x\tan(\phi - \delta) \tag{7.5}$$

where γ is the angle of the powder level with the first flight segment OA (see Figure 7.2).

Its intersection with the line tracing the second segment has the following abscissa:

$$x_2 = \frac{a_2}{\tan(\gamma) - b_2} \tag{7.6}$$

with

$$y_2 = a_2 + b_2 x_2 \tag{7.7}$$

as ordinate.

The intersection of the solid level line with the drum wall has the following abscissa:

$$x_w = -\frac{B_w \pm \sqrt{B_w^2 - 4A_w B_w}}{2A_w} \tag{7.8}$$

with $A_w = 1 + [\tan(\alpha)]^2$, $B_w = 2X_o[\cos(\alpha) - \tan(\gamma)\text{sen}(\delta)] + 2y_o[\tan(\gamma)\cos(\delta) + \text{sen}(\alpha)]$ and $B_w = Ro^2 + R^2$, and its ordinate is given by

$$y_w = x_w \tan(\gamma) \tag{7.9}$$

Three types of powder fills can occur:

- The particles reach the dryer wall. This will occur if $\gamma > arc\ \tan\left(\frac{y_B}{x_B}\right)$, since the section area occupied by the particles is given by

$$S = \frac{R^2}{2}[\beta - sen(\beta)] + \frac{1}{2}\left|x_A y_B - x_B y_A + x_B y_w - x_w y_B\right| \tag{7.10}$$

  with $\beta = 2arcsen\left[\frac{\sqrt{(x_B - x_w)^2 + (y_B - y_W)^2}}{2R}\right]$.

- The particles that do not reach the wall, but reach the second segment. This will happen if $\gamma < arc\ \tan\left(\frac{y_B}{x_B}\right)$, $\sqrt{(x_2 - x_A)^2 + (y_2 - y_A)^2} < l'$ and $y_2 > 0$, since the area of the transversal section occupied by the material is given by

$$S = \frac{1}{2}\left|x_A y_2\right| \tag{7.11}$$

- The flights are empty. This will happen if:

$$y_2 < 0 \tag{7.12}$$

The ratio between the area occupied by the solids in the flights (S) and the load of solids in the flights (h*) may be given by the following equation:

$$h^*(\theta_i) = S_i L \rho_s \tag{7.13}$$

Glikin (1978) proposed the following equation for calculation of the height of falling curtains in rotary drums, which is the straight-line distance between

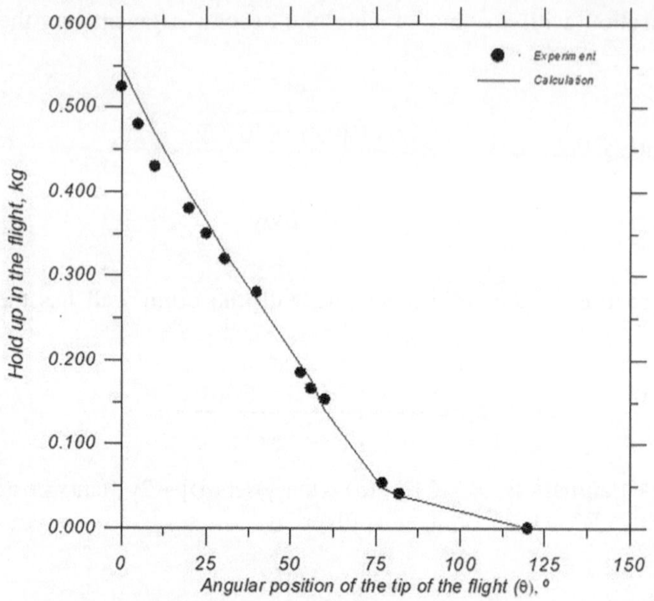

FIGURE 7.3 Variation in solids holdup in the flight with angular position of the flight tip.

the flight tip, where the fall begins, and the particle bed in the bottom part of the dryer:

$$Y_d = \frac{Y_o + \sqrt{R^2 - X_o^2}}{\cos \alpha} \tag{7.14}$$

where $Y_o = R_0 \cos \theta$ and $X_o = R_0 \text{sen}\theta$.

Lisboa et al. (2007) used these equations to predict the solids' holdup in the flights of a rotary dryer applied to fertilizer drying. Figure 7.3 shows the comparison between the holdup measured experimentally for each flight's angular position and the results obtained with the proposed equations (Eq. 7.2 to Eq. 7.14) for the prediction of solids' holdup for the two-segment flights. Figure 7.4 shows the same for the height of falling curtains. The results in Figures 7.3 and 7.4 show that for both the equations, for solids' holdup in the flights and the height of falling curtains, provided accurate estimates of these parameters for this flight configuration.

7.2.3 Design Loading of a Flighted Rotary Dryer

There are three potential degrees of loading in rotary dryers: underloaded, design loaded (optimum loading), or overloaded. However, the drum loading is directly related to the flight loading. Thus, this classification is related to the first unloading flight (FUF), which is the flight that first discharges solids (Karali et al., 2015;

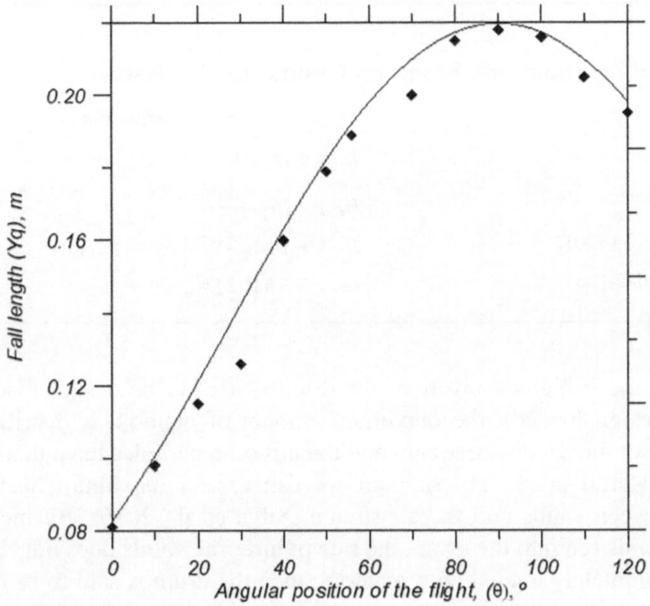

FIGURE 7.4 Variation in the average height of falling curtains in the flight with angular position.

Nascimento et al., 2015). If the flight discharge starts when the flight tip is in the upper half of the drum, after the 9 o'clock position, which can be observed in Figure 7.5, the device is considered underloaded. In this case, the flight holds less material than its capacity, and the time that particles spend in the airborne phase is minimum, which can lead to shorter residence times than the required periods. Therefore, the dryer will be operating below its capacity, i.e., inefficiently. If the holdup of the drum increases, the first unloading position of the flight (FUF) decreases, because the flights will be filled with more material and the discharge of the solids will start at lower angular positions. If unloading starts when the flight tip

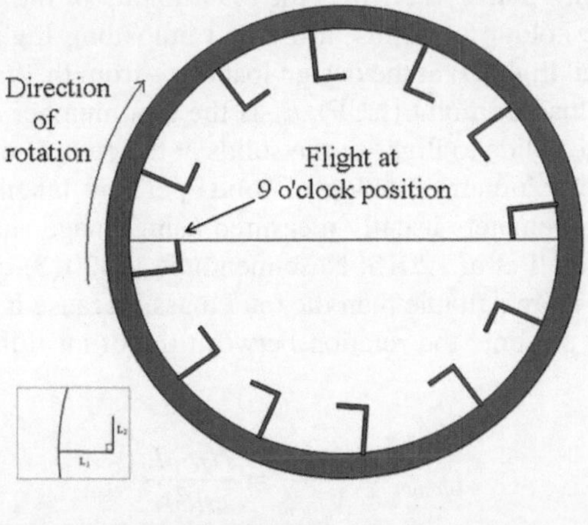

FIGURE 7.5 Flight at the 9 o'clock position.

TABLE 7.1
Design Loading Equations Based on Geometric Analysis

Authors	Equation	
Porter (1963)	$H_{TOT} = H_{FUF}\frac{n_F}{2}$	(7.15)
Kelly and O'Donnel (1977)	$H_{TOT} = H_{FUF}\frac{n_F+1}{2}$	(7.16)
Ajayi and Sheehan (2012)	$H_{TOT} = 1.24((2\sum_{FUF}^{LUF} H_i) - H_{FUF})(1 + Y)$	(7.17)
Karali et al. (2015)	$H_{TOT} = 1.38((2\sum_{FUF}^{LUF} H_i) - H_{FUF})$	(7.18)

is exactly at the 9 o'clock position, the drum is said to be design loaded. In this condition (design loaded), the maximum amount of material is distributed in the airborne phase: the drying airstream and the airborne particles have their maximum interaction (Karali et al., 2015); then we can expect maximum heat and mass transfers between solids and the air stream (Silva et al., 2016). By increasing the amount of solids fed into the drum, the rate of airborne solids does not rise, because flights are completely loaded with material; then the drum is said to be overloaded. This loading condition results in a proportion of material transported by kiln action, and the contact with hot air is limited (Silverio et al., 2014).

Some studies in literature have used geometric models based on the cross-section of flights to estimate the design loading of a rotating drum. Table 7.1 summarizes these geometric models. The model proposed by Porter (1963) (Eq. 7.15) is based on the concept that, at design loading, there are sufficient solids to fill half of all flights. Kelly and O'Donnell (1977) (Eq. 7.16) used an equal angular distribution in the configuration of the flights, which was different to that of Porter (1963). Ajayi and Sheehan (2012) proposed Eq. 7.17 based on the Baker (1988) equation using a new parameter: Y, that is, the ratio of gas-borne solid to flight-borne solid at design loading. Karali et al. (2015) proposed a similar equation with a new fitting factor (Eq. 7.18). However, these authors neglected the Y factor, because it presented low values, within a typical range of 0.042–0.078.

In the equations in Table 7.1, H_{TOT} is the total holdup of the drum at the design loading, H_{FUF} is the holdup of solids in the first unloading flight (FUF), H_i is the holdup in each loaded flight (i) at the design loading – from the first unloading flight (FUF) to the last unloading flight (LUF), n_F is the total number of flights, and Y is the ratio of gas-borne solids to flight-borne solids at the design loading. H_{Tot}, H_{FUF}, and H_i represent the volumetric holdup of solids per unit length of the drum and have units of area (parameters usually measured using image analysis techniques).

Some authors (Karali et al., 2015; Nascimento et al., 2015) consider the use of the filling degree (f) more suitable than the total mass, because it is a dimensionless parameter. Eq. 7.19 presents the relation between the drum filling degree and the total solids holdup.

$$f_{design} = \frac{V_{TOT}}{V_{drum}} = \frac{H_{TOT}L}{\pi R^2 L} \tag{7.19}$$

V_{TOT} is the total volume of solids, V_{drum} is the total volume of the drum, R is the drum radius, and L is the drum length.

The flight cascading rate strongly affects the heat and mass transfer, and varying the Froude number ($Fr = \omega^2 R/g$) of the drum can control this rate (Sunkara et al., 2013). Therefore, the rotational speed has a significant effect on the performance of rotary dryers. Higher rotational speeds increase the discharge of solids to the airborne phase, which increases the total holdup for a design loaded drum (Ajayi and Sheehan, 2012). However, although the rotational speed is implicit in the solids mass in the flights – variable usually used to calculate the design loading – the literature models do not estimate the rotational speed for design loading condition as a function of the drum loading. Thus, it is important to consider this variable in the design loading models.

The FUF usually estimates the design loading condition (Eqs. 7.15–7.19). However, the LUF is also important, because it is the flight in the angular position where the discharge ends (Arruda et al., 2009). With higher LUF angular positions, flights transport particles for longer distances, and the area of solids in the airborne phase is larger, which improves the drying rate (Liu et al., 2005). However, the rotational speed of the drum affects the LUF; thus, the influence of this operating variable must be considered in the design loading predictions.

Nascimento et al. (2018) investigated the design-loaded condition of a flighted rotating drum and for the first time in literature analyzed its dependency on the rotational speed. Image analysis technique was used by these authors to determine the design loading of a rotary drum. Different solid materials and rotational speeds were used to analyze the effect of the rotational speed on the last unloading flight (LUF). A model was developed by these authors that related the Froude number and the physical properties of particles with the filling degree of the drum.

Table 7.2 presents the experimental results of the design loading obtained by Nascimento et al. (2018) as well as the deviation between the geometric model (Eqs. 7.15–7.18) predictions and the experimental results. A drum filling degree ranging from 10.5% to 15.5% was used in their experiments. The used rotational speeds were the ones for the drum be operating at design loading condition for these filling degrees. Nascimento et al. (2018) observed that all geometric models underestimated the design loading. The higher deviations were obtained with Porter (1963) and Kelly and O'Donnell (1977) models, with an average deviation of 34,0% and 21,2%, respectively. The best prediction was obtained with the Karali et al. (2015) model (Eq. 7.18), with an average deviation of 13.3%. However, to improve the fit by this geometric model, a new parameter was estimated (1.59) by Nascimento et al. (2018) using their experimental data, and thus a new equation (Eq. 7.20) was obtained. The prediction using this modified model (Eq. 7.20) presented the lowest deviation from their experimental results (9.2%).

$$H_{TOT} = 1.59\left(\left(2\sum_{FUF}^{LUF} H_i\right) - H_{FUF}\right) \tag{7.20}$$

Karali et al. (2015) proposed their model (Eq. 7.18) based on experimental data obtained from a rotary drum operating with the same number of flights and the same

TABLE 7.2
Prediction of the Design-Loading Models: Percentage of Deviation from Experimental Results

Material	Rotational Speed (rpm)	Experimental Design Loading	Percentage of Deviation from Experimental Results				
			Porter (1963)	Kelly and O'Donnel. (1977)	Ajayi and Sheehan (2012)	Karali et al. (2015)	Nascimento et al. (2018)
Filter sand	22.8	10.5%	−31.6%	−25.9%	−13.3%	−3.6%	10.5%
Glass beads	23.5	10.5%	−37.3%	−32.1%	−21.1%	−12.2%	1.0%
Granulated sugar	12.2	10.5%	−20.3%	−13.7%	−6.1%	4.5%	20.0%
Filter sand	43.8	13%	−43.0%	−38.3%	−24.4%	−15.9%	−3.1%
Glass beads	38.7	13%	−45.5%	−41.0%	−33.2%	−25.6%	−10.0%
Granulated sugar	34.8	13%	−33.3%	−27.7%	−11.1%	−1.1%	12.3%
Filter sand	64.2	15.5%	−51.6%	−47.6%	−32.4%	−24.8%	−13.5%
Glass beads	61.4	15.5%	−48.0%	−43.7%	−30.4%	−22.6%	−11.0%
Granulated sugar	54.5	15.5%	−41.0%	−36.1%	−18.7%	−9.6%	1.9%

shape (L-shaped) of the Nascimento et al. (2018) study. Thus, these results show the limitations of predicting the design loading using geometric models. These models were developed for a particular number and type of flights and have a limited applicability.

As geometric models are not suitable for predicting the design loading and do not consider the effect of rotation speed on this parameter, Nascimento et al. (2018) proposed a new empirical model, which included the influence of the rotational speed. However, in this new model, they used the Froude number, because it is a dimensionless parameter. In addition, some important parameters related of physical and flowability properties of the materials (μ,ρ and σ) were also included in their proposed model, observing the effect of each variable in the Froude number (Fr) for design loading. Eq. 7.21 presents the equation proposed by Nascimento et al. (2018) to predict the Froude number for the design loading as a function of the filling degree and the physical properties of the materials.

$$Fr = \frac{15}{\mu} f^2 - 6.5 \times 10^4 \frac{1}{\rho\sigma} f - 0.103 \tag{7.21}$$

where σ is the static angle of repose, μ is the coefficient of dynamic friction, ρ is the particle density, and f is the filling degree.

In Eq. 7.21, the Froude number to operate at design loading is lower for the materials with higher dynamic friction coefficients. This is due to that in this condition, there is the necessity of a lower rotational speed to balance the forces that maintain particles in the flights (centrifugal, gravitational, and frictional forces). In addition, the influence of particle density on the Froude number to operate at design loading was also significant. Therefore, identifying the material properties and the adequate rotational speed is crucial to predict the design loading condition.

Figure 7.6 shows the prediction by Eq. 7.21 and the experimental data of Nascimento et al. (2018). The predicted values presented a good agreement with the experimental results for all materials used in their study. The maximum deviation from the experimental results was 4.4%, and this value is smaller than the ones observed by the geometric model fit in this work.

The last unloading flight (LUF) position as a function of the rotational speed was also determined by Nascimento et al. (2018). They observed that the LUF position is strongly dependent on the rotational speed. With higher rotational speeds, the discharge of the flights finished at higher angular positions, because the centrifugal force increased and the flights carried particles for longer distances. Therefore, with higher rotational speeds, the drying surface area (active region) in the drum was larger. In addition, Nascimento et al. (2018) observed that the LUF position for filter sand and granulated sugar was larger than the one observed for glass beads. Glass beads presented lower dynamic friction coefficients and hence lower repose angles. Thus, for the same rotational speed, glass beads had a better flowability than the other particles analyzed by these authors. Nascimento et al. (2018) proposed a

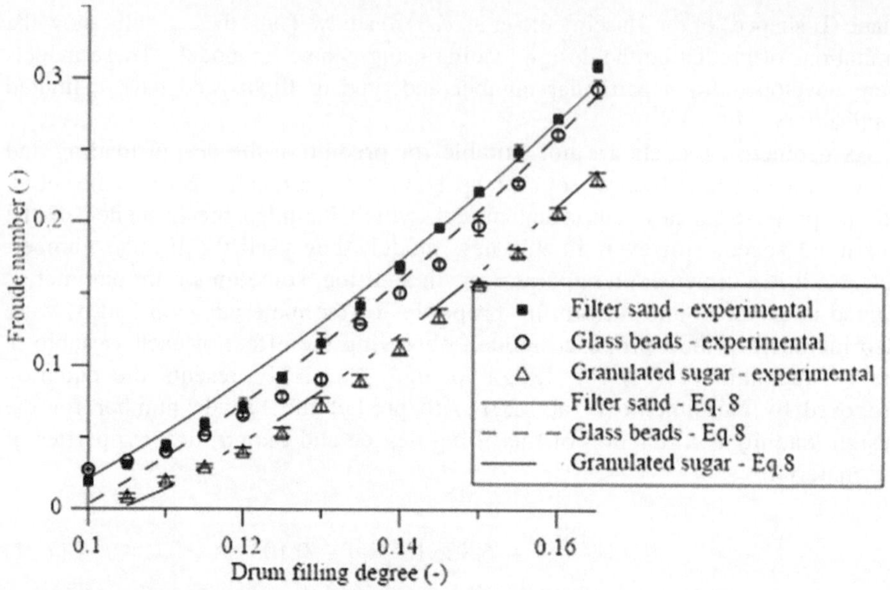

FIGURE 7.6 Experimental and calculated (Eq. 7.21) Froude number as a function of the filling degree (Nascimento et al. 2018).

new equation (Eq. 7.22) to estimate the LUF position as a function of the rotational speed of the drum (Froude number):

$$LUF\,(°) = 206.9\frac{1}{\gamma}\,\ln(Fr) + 169.7 \tag{7.22}$$

Figure 7.7 shows the experimental data and the results predicted by Eq. 7.22 (Nascimento et al., 2018), in which it can be seen that the prediction by Eq. 7.22 presented a good agreement with experimental results for all the materials analyzed by these authors.

7.2.4 CFD Simulations of a Flighted Rotary Drum

The numerical simulations have become an important tool in optimizing the design and the operation of industrial processes (Barrozo et al., 2006). The two-fluid model (Euler-Euler approach) along with the kinetic theory of granular flow (KTGF) and the discrete element method (DEM) are the most used methodologies in simulating granular flows (Santos et al., 2013).

In DEM, each particle trajectory is calculated based on all active forces, providing information about particles at the microscopic level. Although many researchers have adopted this kind of method in the simulations of flighted rotating drums (Geng et al., 2009; Silverio et al., 2014), this methodology has a crucial problem – its computational cost, which is directly proportional to the number of particles, making it difficult to simulate large-scale equipment. In addition, this approach also requires many parameters to describe the process.

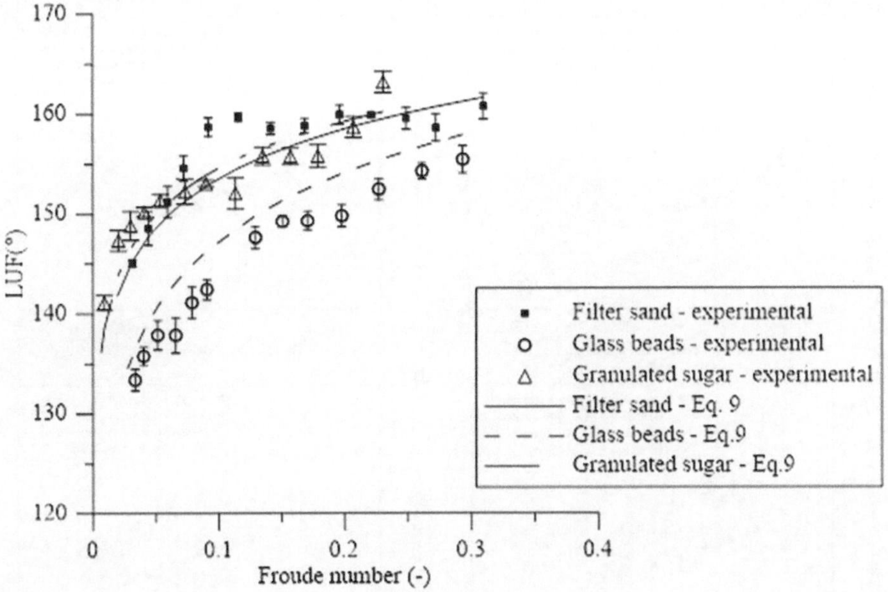

FIGURE 7.7 Experimental and calculated (Eq. 7.22) LUF angular position as a function of the Froude number (Nascimento et al. 2018).

The Euler–Euler approach treats all the phases mathematically as continuous and interpenetrating. The properties of particles are calculated based on the kinetic theory of granular flow. Although DEM simulations provide detailed information about particle flow, Euler-Euler simulations are best suited for large-scale granular flow. As compared with DEM, in the Eulerian approach fewer parameters must be defined.

Numerical simulation studies using the Euler–Euler approach along with the KTGF have become popular in several applications (Cunha et al., 2009; Barrozo et al., 2010; Santos et al., 2009; Oliveira et al., 2009; Pereira et al., 2007; Santos et al., 2012). However, there are a restricted number of studies related to the fluid dynamic study of flighted rotating drums using this methodology. Wardjiman et al. (2008) and Ajayi and Sheehan (2012) studied a curtain of particles that fall through an airstream. Nascimento et al. (2015) and Machado et al. (2017) used the Euler-Euler approach to study a flighted rotating drum. The authors analyzed the solids discharge from the flights by varying the rotational speed. These previous works (Nascimento et al., 2015 and Machado et al., 2017) were performed in a bench scale and the rotational speeds varied from 3 to 5 rpm, that are the typical rotational speeds used in industrial equipment. However, according to Perry and Green (1999), rotary dryers must operate at peripheral speeds of 0.1–0.5 m/s. Thus, in the bench scale, the rotational speed should be higher than the values used by most authors in the literature (Fernandes et al., 2009).

Nascimento et al. (2019) performed a study of solids fluid dynamics for a flighted rotary drum operating at high rotational speeds (21.3 and 36.1 rpm), that lead to peripheral speeds of 0.12 and 0.20 m/s, which is in the range recommended

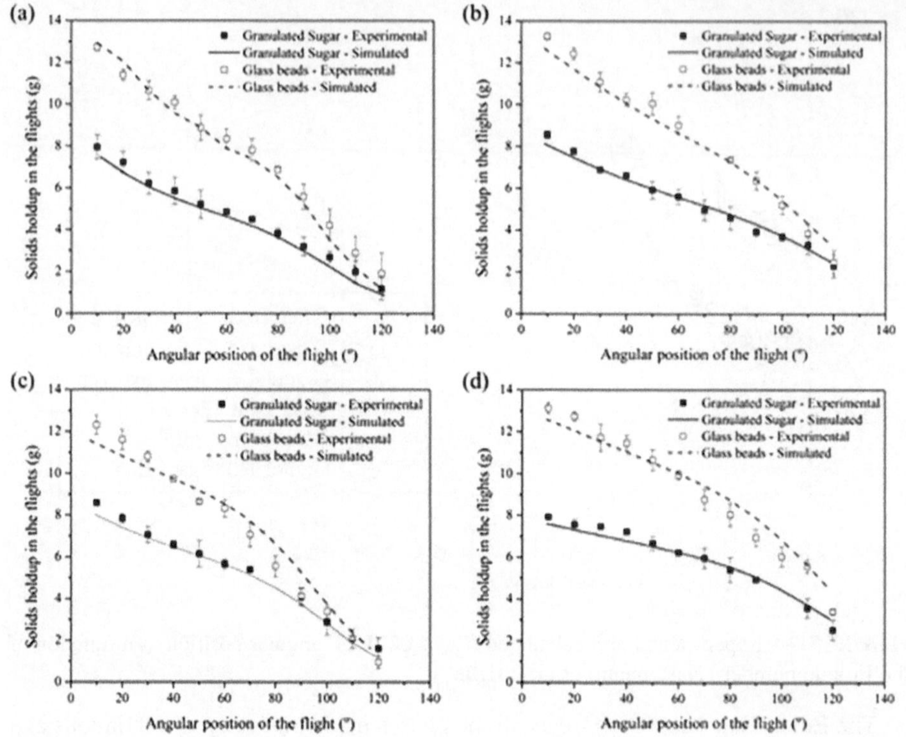

FIGURE 7.8 Comparison between simulation and experimental data for the drum operating with glass beads and granulated sugar: (a) 12 flights and 0.027 Froude number, (b) 12 flights and 0.078 Froude number, (c) 15 flights and 0.027 Froude number, (d) 15 flights and 0.078 Froude number (Nascimento et al. 2019).

by Perry and Green (1999). These authors used an Eulerian approach to predict the flight's discharge in a rotary drum. The effect of some operating conditions (rotation speeds and the number of flights) on the fluid dynamic behavior of different solid materials (glass beads and granulated sugar) in a flighted rotating drum was analyzed. The simulated results were compared with experimental data obtained in a bench scale rotary drum. Figure 7.8 presents the simulated and experimental results of the solids' holdup in the flight as a function of the angular position for the drum operating with glass beads and granulated sugar at Froude numbers of 0.027 and 0.078 for the rotary drum operating with 12 and 15 flights (Nascimento et al., 2019).

The results of Figure 7.8 show that solids' holdup in the flights increased with an increase in rotational speed for all angular positions and for both numbers of flights, leading to an enhancement of contact between solids and air. The angular position of the last unloading flight (LUF) is also strongly dependent on the rotational speed. With higher rotational speeds, the discharge of the flights finishes at higher angular positions, because the centrifugal force increases and the flights carry particles for longer distances. Therefore, with increased Froude numbers, the solids' holdup in the flight and the LUF are also increased; thus, the drying surface area (active region) in the drum is also larger. It also can be seen that the mass of glass beads in the flights is always greater than that for the granulated sugar at the same angular position. A greater bulk density of glass beads is responsible for this behavior.

7.3 ROTARY DRUMS WITHOUT FLIGHTS

7.3.1 Introduction

Apart from the usage of flighted rotary drums, rotary drums without flights or unbaffled rotary drums are applied in many industrial processes such as drying, granulation, mixing, and coating (Mellmann, 2001; Gerasimov and Volkov, 2015; Tada et al., 2020). According to Mellmann (2001), depending on the drum geometric dimensions and operating conditions, and particle properties, different granular flow regimes take place inside a rotary drum without flights, i.e., sliding, slumping, rolling, cascading, cataracting, and centrifuging regimes. Table 7.3 summarizes the granular flow regimes as functions of the Froude number ($Fr = \omega^2 R/g$), the drum-filling degree (f), and the drum-wall friction coefficient (μ_w), along with their main applications.

The sliding and slumping regimes can occur at low rotation speeds and filling degrees combined with smooth drum walls. Under the sliding regime, the bed material slides adjacent to the drum wall and no particle mixing is observed, whereas in the slumping regime the granular material moves as a rigid body reaching an upper angle of repose and then avalanches until a lower angle of repose is established. The rolling regime takes place at higher rotation speeds where the bed material presents a flat surface with a constant dynamic angle of repose. Further speed increments can curve the bed material surface, like an "S"-shaped curve, which corresponds to the cascading regime. Under even higher rotation speeds, the cataracting regime can show up, where particles are projected from the bed material surface into the air space. For $Fr \geq 1$, the particles fare pushed against the drum wall forming a ring all together (centrifuging regime). The sliding and centrifuging regimes are not used at all and must be avoided (Blumberg and Schlünder, 1996; Mellmann 2001; Santos et al., 2013; Santos et al., 2016a).

The rolling regime has been extensively studied due its broad applications (Santos et al., 2016b; Benedito et al., 2018). Two different regions can be observed for the drum operating under the rolling regime: the passive region (near the wall), where particles move as a solid body (lower particle velocities), and the active region (near the bed surface), where the particles avalanche and cascade downward (higher particle velocities). The interface between the two regions presents a velocity inflexion point where a reverse of the flow takes place, which can be observed in Figure 7.9.

The physical mechanisms, such as mixing and segregation, and heat and mass transfer, mainly occur in the active region (Ding et al., 2001; Dubé et al., 2013). According to Santos et al., the higher the drum-filling degree or the drum rotation speed, the higher the active region thickness. Furthermore, the authors found that the particle velocity profile in the active region was not parabolic but rather linear at the mid-chord position (Reference line in Figure 7.9), which is in accordance with those finds reported by Boateng and Barr (1997).

TABLE 7.3
Different Granular Flow Regimes in a Rotary Drum Without Flights (Mellmann, 2001)

Granular Flow Regimes					
Sliding regime	**Slumping Regime**	**Rolling Regime**	**Cascading Regime**	**Cataracting Regime**	**Centrifuging Regime**
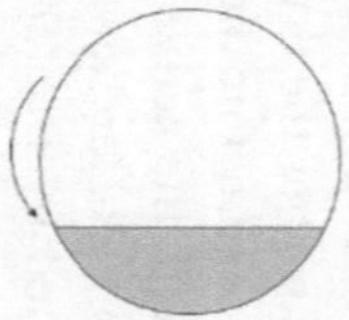	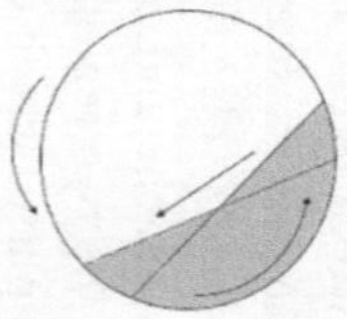	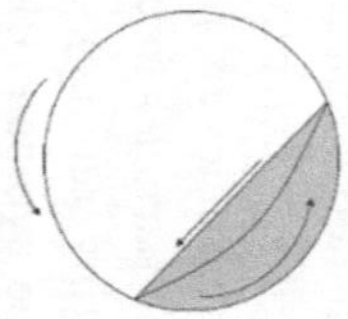	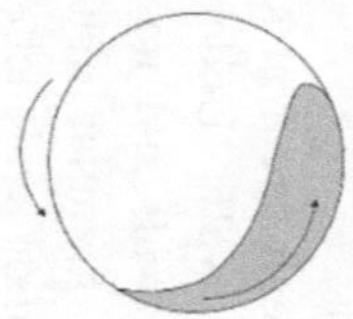	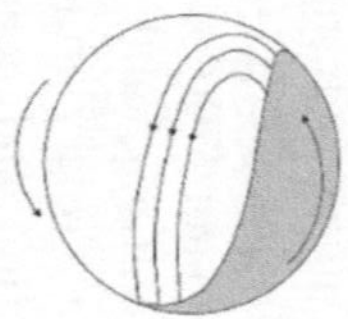	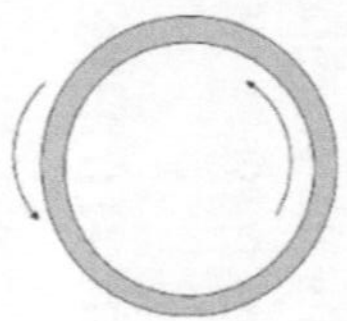
Range of Froude number, Fr [-]					
$0 < \mathrm{Fr} < 10^{-4}$	$10^{-5} < \mathrm{Fr} < 10^{-3}$	$10^{-4} < \mathrm{Fr} < 10^{-2}$	$10^{-3} < \mathrm{Fr} < 10^{-1}$	$0.1 < \mathrm{Fr} < 1$	$\mathrm{Fr} \geq 1$
Range of filling degree, f[-]					
$f < 0.1$	$f < 0.1$	$f > 0.1$	$f > 0.1$	$f > 0.2$	$f > 0.2$
Wall friction coefficient, μ_w [-]					
$\mu_w < \mu_{w,c}$	$\mu_w > \mu_{w,c}$	$\mu_w > \mu_{w,c}$	$\mu_w > \mu_{w,c}$	$\mu_w > \mu_{w,c}$	$\mu_w > \mu_{w,c}$
Application					
no use	Rotary kilns and reactors; rotary dryers and coolers; mixing drums			Ball mills	no use

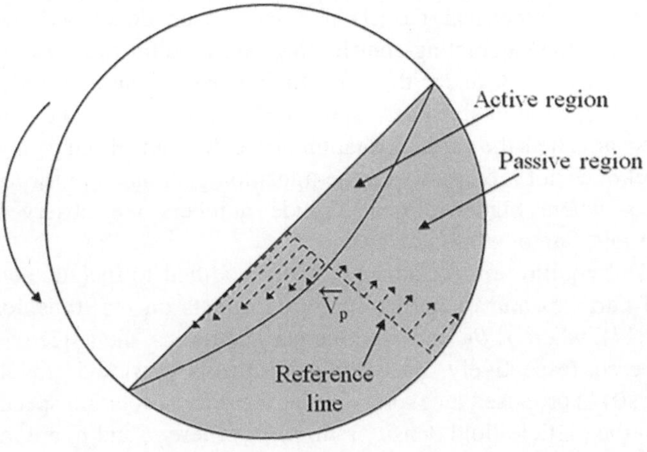

FIGURE 7.9 Active and Passive regions in the rolling regime.

7.3.2 Prediction of Granular Flow Regimes in a Rotary Drum without Flights

The efficiency in different industrial processes depends on the granular flow regime (Barrozo et al., 1998). Thus, the prediction of the particle motion inside a rotary drum, including the effect of the granular material properties, is of fundamental importance. Most of these flow regimes have been studied, although there are a restricted number of works dedicated to the transition phenomenon between the regimes (Henein et al., 1983; Watanabe, 1999; Mellmann, 2001; Santos et al., 2016a).

Herein, a special focus is placed on the cataracting-centrifuging transition, since all the other regimes are characterized in a dimension-less way based on a percentage of the critical Froude number for centrifuging. Table 7.4 summarizes the

TABLE 7.4
Theoretical and Semi-Empirical Models for the Cataracting-Centrifuging Transition

Authors	Models	
Classical mechanics	$\omega_c = \sqrt{\frac{g}{R}}$	(7.23)
Rose and Sullivam (1957)	$\omega_c = \sqrt{\frac{2g}{2R-2r}}$	(7.24)
Walton and Braun (1993)	$\omega_c = \sqrt{\frac{g}{Rsin\theta_s}}$	(7.25)
Ristow (1998)	$\omega_c = \sqrt{\frac{g}{R\sqrt{1-f}}}$	(7.26)
Watanabe (1999)	$\omega_c = \sqrt{\frac{g}{Rsin\theta_s\sqrt{1-f}}}$	(7.27)
Juarez et al. (2011)	$\omega_c = \sqrt{\frac{g\left(1-\rho_f/\rho_s\right)}{Rsin\theta_s\sqrt{1-f}}}$	(7.28)

theoretical and simi-empirical models that have been developed over the past decades to predict the cataracting-centrifuging regime transition (Table 7.4).

According to Eq. 7.23, under the centrifuging regime, an equilibrium between centrifugal and gravitational forces is reached (Fr = 1) and the corresponding rotation speed is called the critical rotation speed for centrifuging (ω_c). This theoretical model does not represent well the transition phenomenon for low values of filling degree, where higher critical Froude numbers are observed (Fr > 1) (Watanabe, 1999; Santos et al., 2016a).

The classical equilibrium formula has been modified to include some physical properties of particles and operating condition effects on the transition modeling (Eqs. 7.24–7.27), where r, θ_S, and f are the particle radius, the angle of repose, and the filling degree, respectively. Based on experiments in wetted granular systems, Juarez et al. (2011) proposed an expression for the critical rotation speed taking into consideration the particle-fluid density ratio ρ_f/ρ_s, where ρ_f and ρ_s are, respectively, the densities of the fluid and solid phases (Eq. 7.28).

Santos et al. (2016a) have recently proposed a semi-empirical model to predict the cataracting-centrifuging transition using particles of different densities, diameters, and shapes (Eq. 7.29), by introducing two additional parameters (λ and τ). In their model, λ represents the additional effect of the critical Froude number under low values of the filling degree, and τ denotes the exponential decay of the critical rotation speed (ω_c) with increasing the filling degree.

$$\omega_c = \sqrt{\frac{g[1 + \lambda \exp(-\tau f)]}{R}} \tag{7.29}$$

For high values of filling degree ($f \to 1$), Eq. 7.29 tends to the classical equilibrium formula (Eq. 7.23), whereas for low values of filling degree ($f \to 0$) the critical rotation speed for centrifuging, as well as the critical Froude number exponentially increases. This behavior is in accordance with experimental observations (Watanabe, 1999; Mellmann, 2001; Santos et al., 2016a).

The authors also observed a hysteresis phenomenon during the centrifuging transition, i.e., the transition behavior by increasing the drum rotation speed (increasing curve) was shown to be different from the corresponding one by decreasing the drum rotation speed (decreasing curve). Table 7.5 shows the adjusted parameters from Eq. 7.29 for different granular materials (Group 1: irregular particles; Group 2: rounded particles).

The hysteresis effect was strong in the case of rounded particles and almost negligible for a flattened particle (Table 7.5). Furthermore, the critical rotation speeds for centrifuging in the case of irregular particles were lower than the ones for rounded particles under low values of filling degree (Santos et al., 2016a).

7.3.3 Particle Segregation in a Rotary Drum without Flights

The particle segregation phenomenon takes place in industrial processes involving multi-component heterogeneous granular materials, e.g., particles with differences

TABLE 7.5
Experimental Adjusted λ and τ Parameters Through Eq. 7.29 for Different Granular Materials (Santos et al., 2016a)

Granular materials		Increasing Curve*		Decreasing Curve*	
		λ_I	τ_I	λ_D	τ_D
Group 1	Tablet	0.88	2.34	0.66	2.37
	Corn	1.25	3.33	0.96	3.30
	Rice	0.97	2.89	0.80	2.88
Group 2	Soybean	4.40	6.66	1.49	6.18
	Glass beads (d_p = 1.13 mm)	3.24	6.00	1.15	3.64
	Glass beads (d_p = 4.22 mm)	8.38	8.35	1.23	6.44

Note
* The correlation coefficients were above 0.96 for all the cases.

in size, density, shape, and surface roughness (Fan et al., 1970; Ottino and Khakhar, 2000). Particle segregation is shown to negatively affect the process performance in rotary drums, where minor differences in physical properties can segregate particles when rotary drums are in movement, yielding poor mixing zones. Two different categories of segregation can be identified in rotary drums without flights, as depicted in Figure 7.10: the radial segregation, where smaller (percolation mechanism) or denser (buoyancy mechanism) particles migrate towards the center of the drum forming a transversal core; and the axial segregation, where particles form alternating bands along the drum axial direction (Williams, 1976).

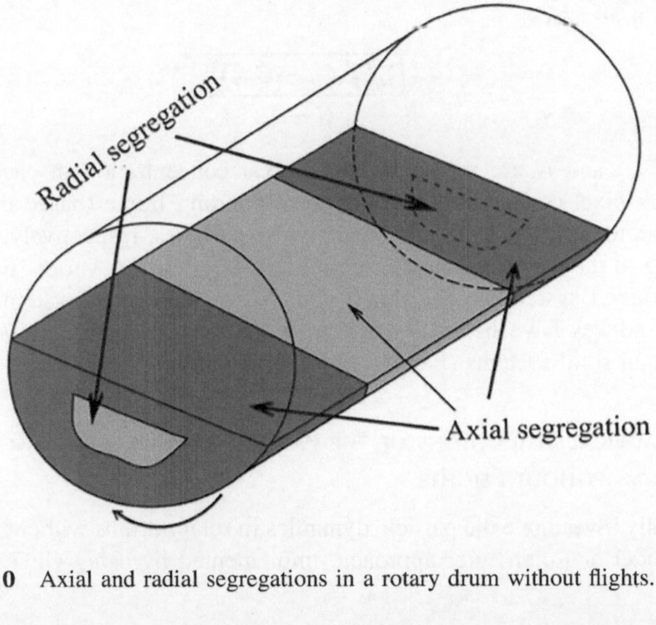

FIGURE 7.10 Axial and radial segregations in a rotary drum without flights.

The radial and axial segregations in rotary drums have been widely investigated (Liu et al., 2015; Santos et al., 2016c). Jain et al. (2005) reported an interesting study concerning the segregation in a rotary drum by combining particles of different sizes and densities. According to the authors, a mixture composed of particles whose diameter ratio was greater than the density ratio inhibited the segregation process. Lee et al. (2013) investigated the effects of the drum end-caps roughness on the particle segregation by using two different drum configurations: end-cap walls with similar roughness (symmetric) and end-cap walls with different roughness (asymmetric). The authors claimed that the particle size ratio and the drum geometric parameters play an important role in the axial particle segregation. An asymmetric axial band, composed of smaller particles, was formed in the case of end-cap walls with different roughness (Figure 7.10).

Santos et al. (2016c) applied image treatment analysis to assess the effects of the rotary drum equipment itself, the particle diameter, the particle density, and the initial particle loading on particle segregation. Two different initial particle loadings were used, i.e., side-side and top-bottom loadings. To analyze the segregation promoted exclusively by the equipment, a monodisperse system, comprised of equal proportions of black-dyed and white-dyed particles, was used. For the image analysis, the authors followed the steps below:

- after recording the particle movement under different conditions, digital still frames were isolated from each video;
- the resulting frames were then converted to 8-bit grayscale, and filtered to produce just black and white pixels;
- each black-and-white frame was divided into equal-sized cells (mixture cells), and the black pixel concentration was measured in each cell.

The following segregation index (I_S in Eq. 7.30) was used to quantify the segregation level in all cases:

$$I_S = \sqrt{\frac{\Sigma_{i=1}^{N} (C_b - C_{avg})^2}{N - 1}} \tag{7.30}$$

where C_b, C_{avg} and N are the local black pixel concentration in each cell, the average black pixel concentration in the corresponding frame (based on the total cells), and the total number of cells occupied by particles, respectively.

According to the segregation index definition (Eq. 7.30), I_S varies from 0, for a completely mixed system, to 0.5, for a completely segregated system. Different segregation indexes have been used in the literature to quantify the segregation phenomenon in solid systems (Fan et al., 1970; Chou et al., 2010).

7.3.4 Numerical Simulations of the Particle Dynamics in a Rotary Drum without Flights

To numerically investigate the particle dynamics in rotary drums without flights, the two-fluid model or Euler-Euler approach, implemented by using CFD techniques

(Computational Fluid Dynamics), and the Lagrangian approach, which is implemented by using the Discrete Element Method (DEM), are commonly applied. These numerical approaches were previously described in Section 7.2.4.

Different granular flow regimes have been successfully described through DEM simulations (Yang et al., 2008; Santos et al., 2016b) and two-fluid model simulations (Santos et al., 2013; Huang et al., 2013; Delele et al., 2016; Liu et al., 2016; Huang et al., 2017). Through a particle collision analysis using DEM, Yang et al. (2008) found that the cascading regime promoted significantly better particle mixing than the slumping and rolling regimes. Santos et al. investigated the influence of different drag force models on the particle velocity distribution in a rotary drum operated in the rolling regime using the two-fluid model. According to the authors, the drag force can be neglected in the case of a rotary drum operated in the rolling regime where there is no fluid entering or leaving the system. Furthermore, the authors numerically assessed the effects of the rotation speed and the filling degree on the active region thickness, which is related to the rates of mass, energy, and momentum transfer.

Good agreements with experimental observations concerning the particle segregation phenomenon have also been achieved through numerical simulations using the two-fluid model (Santos et al., 2016c; Rong et al., 2020) and DEM (Brandão et al., 2020; Hou and Zhao, 2020). The effects of the particle diameter, particle density, drum filling degree, and drum rotation speed on the axial and radial particle segregations were systematically investigated by Brandão et al. (2020) using DEM. The authors experimentally measured the DEM input parameters and the numerical results agreed well against the experimental observations. Qualitative agreements against experimental observations were obtained by Santos et al. (2016c) through CFD (two-fluid model) simulations of the radial segregation, as shown in Figures 7.11 and 7.12,

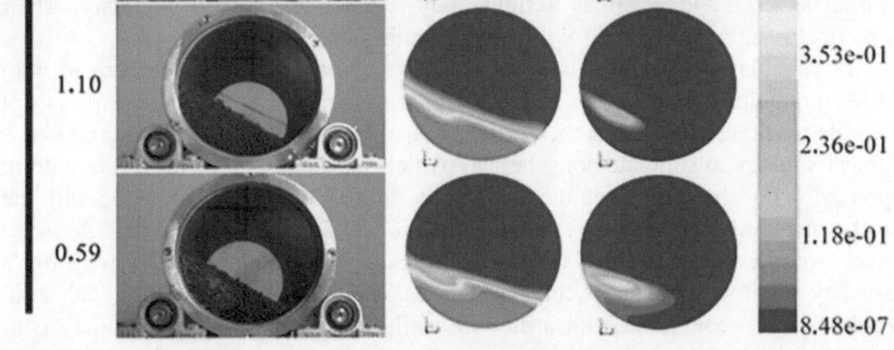

FIGURE 7.11 Radial segregation promoted by the particle size difference (filling degree of 25%): (a) experimental; (b) simulated showing the biggest particles; (c) simulated showing the smallest particles (Santos et al. 2016c).

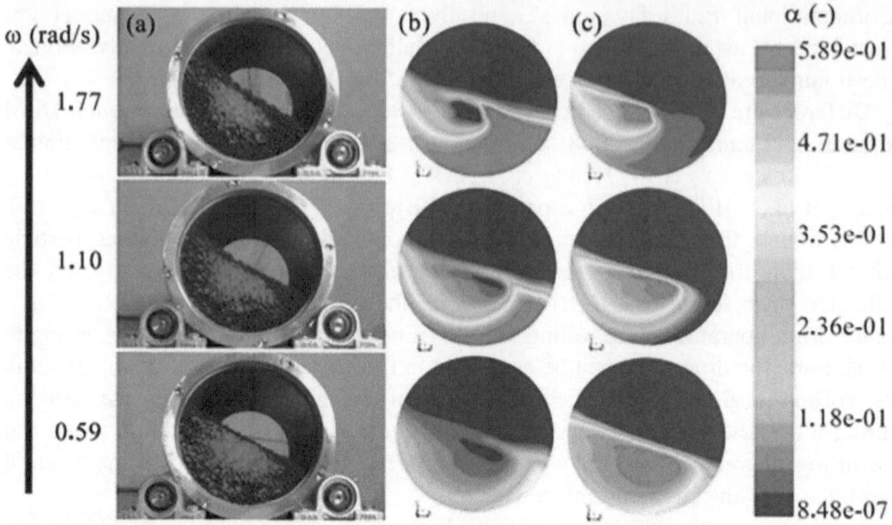

FIGURE 7.12 Radial segregation promoted by the particle size difference (filling degree of 50%): (a) experimental; (b) simulated showing the biggest particles; (c) simulated showing the smallest particles (Santos et al. 2016c).

for the drum operated under different rotation speeds (ω) and filling degrees of 25% and 50%, respectively.

Despite most of DEM and two-fluid model simulations in a rotary drum used spherical particles, the majority of industrial granular materials differ significantly from perfect-shaped spheres. Various attempts to accurately represent the particle shape effects on granular flow simulations in a rotary drum have been done using DEM (Geng et al., 2011; Just et al., 2013; Höhner et al., 2014; He et al., 2019; Jiang et al., 2019). Santos et al. (2016b) applied DEM simulations to investigate the effects of particle shape, drum operating conditions, and drum geometry on the particle velocity distribution and the dynamic angle of repose. For the simulations involving rice grains, the authors used the clump technique that supports the creation of super-particles of arbitrary shape. Each clump consisted of a set of overlapping spheres and acted as a rigid body with a deformable boundary. Their numerical results agreed well with experimental observations.

CFD simulations usually have lower computational costs in comparison with DEM simulations, but because the solid phase is treated as continuous, it is not possible to directly represent the particle shape. Benedito et al. (2018) proposed a pioneer strategy to simulate nonspherical particles (rice grains) inside a rotary drum operated under the rolling regime through the Euler-Euler approach along with the kinetic theory of granular flow and frictional models. In this work, the particle shape effects were proposed to be related to the critical solid fraction by using Schaeffer's viscosity frictional model (Schaeffer, 1987). For rice grains, the critical solid fraction of 0.4, combined with a moving wall method and no-slip boundary condition for the particle velocities at the wall, agreed remarkably well with experiments, as can be seen in Figure 7.13.

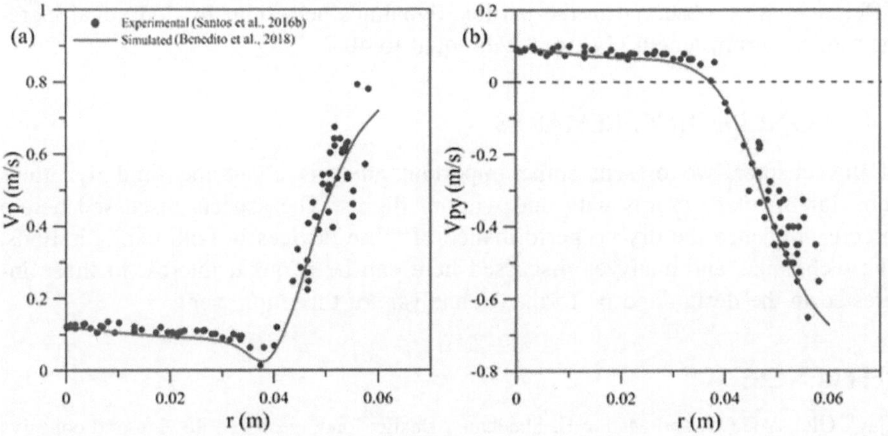

FIGURE 7.13 Experimental (Santos et al. 2016b) and CFD simulation (Benedito et al. 2018) of rice grain velocity distributions at the Reference line shown in Figure7.9: (a) resulting velocity; (b) y-component of velocity.

Benedito et al. (2018) also investigated the drum end-caps' effects on the particle dynamics using the validated model for rice grains. In this case, different values of the ratio between the drum length and the particle diameter (i.e., L/d_s, where L is the drum length and d_s is the particle diameter) were used. As shown in Figure 7.14, no

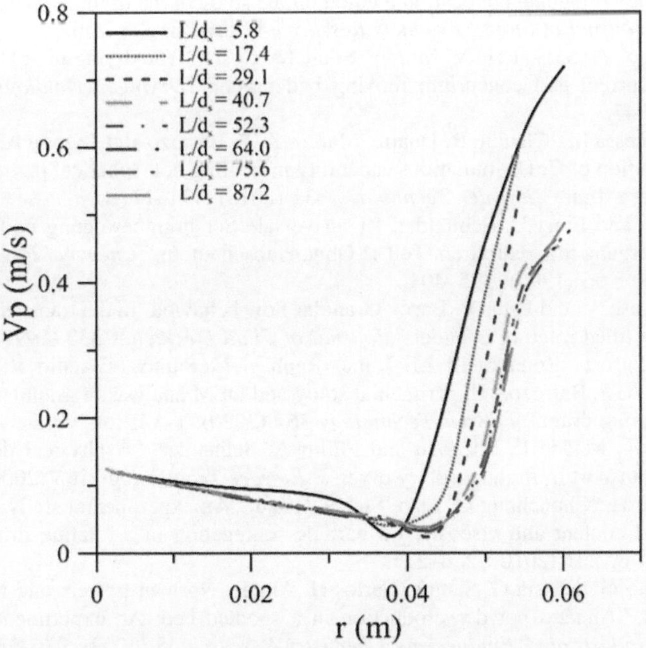

FIGURE 7.14 CFD simulations (Benedito et al. 2018) of rice grain velocity distributions at the Reference line shown in Figure 7.9 for different drum lengths.

differences were observed in the particle dynamics behaviour by additional increment of the drum length (L) from L/d_s equal to 40.7.

7.4 CONCLUDING REMARKS

In this chapter, we present some important analysis about the fluid dynamics behavior in rotary drums with and without flights. The aspects discussed herein directly influence the drying performance of these devices in both configurations. The techniques and analyzes discussed here can be of great interest to those interested in the design and performance analysis of this equipment.

REFERENCES

Ajayi, Olukayode O. and Madoc E. Sheehan. "Design loading of free flowing and cohesive solids in flighted rotary dryers." *Chemical Engineering Science* 73 (2012): 400–411.

Arruda, Edu. B., Juliana M. F. Façanha, Lucas N. Pires, Adilson J. Assis and Marcos A. S. Barrozo. "Conventional and modified rotary dryer: Comparison of performance in fertilizer drying." *Chemical Engineering and Processing* 48 (2009): 1414–1418.

Baker, Christopher G. J. "The design of flights in cascading rotary dryers". *Drying Technology* 6 (1988): 631–653.

Barrozo, Marcos A. S., Claudio R. Duarte, Norman Epstein, N., John R. Grace and Jim C. Lim. "Experimental and computational fluid dynamics study of dense-phase, transition region and dilute-phase spouting." *Industrial & Engineering Chemistry Research* 49 (2010): 5102–5109.

Barrozo, Marcos A. S., Humberto M. Henrique, Demerval J. Sartori and José T. Freire. "The use of the orthogonal collocation method on the study of the drying kinetics of soybean seeds." *Journal of Stored Products Research* 42 (2006): 348–356.

Barrozo, Marcos A. S., Valéria V. Murata, Sônia.M. Costa, "The drying of soybean seeds in countercurrent and concurrent moving bed dryers." *Drying Technology* 16 (1998): 2033–2047.

Benedito, Wanessa M., Claudio R. Duarte, Marcos A. S. Barrozo and Dyrney A. Santos. "An investigation of CFD simulations capability in treating non-spherical particle dynamics in a rotary drum." *Powder Technology* 332 (2018): 171–177.

Blumberg, W. and Ernst U. Schlünder. "Transversale Schüttgutbewegung und konvektiver Stoffübergang in Drehrohren. Teil 1: Ohne Hubschaufeln." *Chemical Engineering and Processing* 35 (1996): 395–404.

Boateng, Akwase A. and Peter V. Barr. "Granular flow behaviour in the transverse plane of a partially filled rotating cylinder." *Journal of Fluid Mechanics* 330 (1997): 233–249.

Brandão, Rodolfo.J., Ronndinelli M. Lima, Raphael L. Santos, Claudio R. Duarte and Marcos A. S. Barrozo. "Experimental study and DEM analysis of granular segregation in a rotating drum." *Powder Technology* 364 (2020): 1–12.

Britton, Paul F., Madoc E. Sheehan and Philip A. Schneider. "A physical description of solids transport in flighted rotary dryers." *Powder Techonology* 165 (2006): 153–160.

Chou, Shihhao H., Chunchung C. Liao, Shu S. Hsiau. "An experimental study on the effect of liquid content and viscosity on particle segregation in a rotating drum." *Powder Technology* 201 (2010): 266–272.

Cunha, Fabiano G., Kassia G. Santos, Carlos H. Ataíde, Norman Epstein and Marcos A. S. Barrozo. "Annatto powder production in a spouted bed: An experimental and CFD study." *Industrial & Engineering Chemistry Research* 48 (2009): 976–982.

Delele, Mulugeta A., Fabian Weigler, George Franke and Jochen Mellmann. "Studying the solids and fluid flow behavior in rotary drums based on a multiphase CFD model." *Powder Technology* 292 (2016): 260–271.

Ding, Yulong L., Jonathan P. K. Seville, Robin Forster, David J. Parker. "Solids motion in rolling mode rotating drums operated at low to medium rotational speeds." *Chemical Engineering Science* 56 (2001): 1769–1780.

Dubé, Olivier, Ebrahim Alizadeh, Jamal Chaouki and François Bertrand. "Dynamics of non-spherical particles in a rotating drum." *Chemical Engineering Science* 101 (2013): 486–502.

Fan, Liangtseng T., S. J. Chen and Colin A. Watson. "Solids mixing." *Industrial & Engineering Chemistry Research* 62 (1970): 53–69.

Felipe, Carlos A. S., Marcos A. S. Barrozo. "Drying of soybean seeds in a concurrent moving bed: Heat and mass transfer and quality analysis." *Drying Technology* 21 (2003): 439–456.

Fernandes, Nilson J., Carlos H. Ataíde and Marcos A. S. Barrozo. "Modeling and experimental study of hydrodynamic and drying characteristics of an industrial rotary dryer." *Brazilian Journal of Chemical Engineering* 26 (2009): 331–341.

Geng, Fan, Yimin Li, Xinyong Wang, Zhulin Yuan, Yaming Yan and Dengshan Luo. "Simulation of dynamic processes on flexible filamentous particles in the transverse section of a rotary dryer and its comparison with video-imaging experiments." *Powder Technology* 207 (2011): 175–182.

Geng, Fan, Zhulin Yuan, Yaming Yan, Dengshan Luo, Hongsheng Wang, Bin Li and Dayong Xu. "Numerical simulation on mixing kinetics of slender particles in a rotary dryer." *Powder Technology* 193 (2009): 50–58.

Gerasimov, Gennady and Eduard Volkov. "Modeling study of oil shale pyrolysis in rotary drum reactor by solid heat carrier." *Fuel Processing Technology* 139 (2015): 108–116.

Glikin, P. G. "Transport of solids through flighted rotation drums." *Transactions of the Institution of Chemical Engineers* 56 (1978): 120–126.

He, Siyuan, Jieqing Gan, David Pinson and Zongyan Zhou. "Particle shape-induced radial segregation of binary mixtures in a rotating drum." *Powder Technology* 341 (2019): 157–166.

Henein, Hani, James K. Brimacombe and Alan P. Watkinson. "Experimental studies of transverse bed motion in rotary kilns." *Metallurgical and Materials Transactions B* 14 (1983): 191–205.

Höhner, Domink, Siegmar Wirtz and Viktor Scherer. "A study on the influence of particle shape and shape approximation on particle mechanics in a rotating drum using the discrete element method." *Powder Technology* 253 (2014): 256–265.

Hou, Zhichao and Yongzhi Zhao. "Numerical and experimental study of radial segregation of bi-disperse particles in a quasi-two-dimensional horizontal rotating drum." *Particuology* 51 (2020): 109–119.

Huang, An-Ni., Wei-Chun Kao and Hsiu-Po Kuo. "Numerical studies of particle segregation in a rotating drum based on Eulerian continuum approach." *Advanced Powder Technology* 24 (2013): 364–372.

Huang, An-Ni and Hsiu-Po Kuo. "CFD simulation of particle segregation in a rotating drum. Part I: Eulerian solid phase kinetic viscosity." *Advanced Powder Technology* 28 (2017): 2094–2101.

Jain, Nitin, Julio M. Ottin and Richard M. Lueptow. "Regimes of segregation and mixing in combined size and density granular systems: An experimental study." *Granular Matter* 7 (2005): 69–81.

Jiang, Shengqiang, Yixuan Ye, Mingxue He, Chunyan Duan, Sisi Liu, Jingang Liu, Xiangwu Xiao, Hao Zhang and Yuanqiang Tan. "Mixing uniformity of irregular sand and gravel materials in a rotating drum with determination of contact model parameters." *Powder Technology* 354 (2019): 377–391.

Juarez, Gabriel, Pengfei Chen and Richard M. Lueptow. "Transition to centrifuging granular flow in rotating tumblers: A modified Froude number." *New Journal of Physics* 13 (2011): 1–12.

Just, Sarah, Gregor Toschkoff, Adrian Funke, Dejan Djuric, Georg Scharrer, Johannes Khinast, Klaus Knop and Peter Kleinebudde. "Experimental analysis of tablet properties for discrete element modeling of an active coating process." *AAPS PharmSciTech* 14 (2013): 402–411.

Karali, Mohamed A., Koteswara R. Sunkara, Fabian Herz and Eckehard Specht. "Experimental analysis of a flighted rotary drum to assess the optimum loading." *Chemical Engineering Science* 138 (2015): 772–779.

Keey, Roger B. *Drying: Principles and Practice* (Oxford: Pergamon Press, 1972).

Kelly, John J. and John P. O'Donnel. "Residence time model for rotary drums." *Transactions of the Institution of Chemical Engineers* 55 (1977): 243–252.

Kemp, Ian C. "Comparison of Particles Motion Correlations For Cascading Rotary Dryers". *Proceedings of the 14th International Drying Symposium (IDS)*, São Paulo, Brazil, B, 790–797, 2004.

Lee, Ching-Fang, Hsien-Ter Chou and Hervé Capart. "Granular segregation in narrow rotational drums with different wall roughness: Symmetrical and asymmetrical patterns." *Powder Technology*. 233 (2013): 103–115.

Lisboa, Michel H., Danilo S. Vitorino, Willian B. Delaiba, José R.D. Finzer and Marcos A. S. Barrozo. "A study of particle motion in rotary dryer." *Brazilian Journal of Chemical Engineering* 24 (2007): 265–374.

Liu, Hong, Hongchao Yin, Ming Zhang, Maozhao Xie and Xi Xi. "Numerical simulation of particle motion and heat transfer in a rotary kiln." *Powder Technology* 287 (2016): 239–247.

Liu, Xiao Y., Eckehard Specht, Jochen Mellmann. "Experimental study of the lower and upper angles of repose of granular materials in rotating drums." *Powder Technology* 154 (2005): 125–131.

Liu, Xiaoyan, Chaoyu Zhang and Jiezi Zhan. "Quantitative comparison of image analysis methods for particle mixing in rotary drums." Powder Technology 282 (2015): 32–36.

Machado, Marcela V. C., Suellen M. Nascimento, Claudio R. Duarte and Marcos A. S. Barrozo. "Boundary conditions effects on the particle dynamic flow in a rotary drum with a single flight." *Powder Technology* 311 (2017): 341–349.

Mellmann, Jochen. "The transverse motion of solids in rotating cylinders-forms of motion and transition behavior." *Powder Technology* 118 (2001): 251–270.

Nascimento, Suellen M., Claudio R. Duarte and Marcos A. S. Barrozo. "Analysis of the design loading in a flighted rotating drum using high rotational speeds." *Drying Technology* 36 (2018): 1200–1208.

Nascimento, Suellen M., Rondinelli M. Lima, Rodolfo J. Brandão, Claudio R. Duarte and Marcos A. S. Barrozo. "Eulerian study of flights discharge in a rotating drum." *The Canadian Journal of Chemical Engineering* 97 (2019): 477–484.

Nascimento, Suellen M., Dyrney A. Santos, Marcos A. S. Barrozo and Claudio R. Duarte. "Solids holdup in flighted rotating drums: An experimental and simulation study." *Powder Technology* 280 (2015): 18–25.

Oliveira, Diogo C., Celso A. K. Almeida, Luís G.M. Vieira, João J.R. Damasceno and Marcos A. S. Barrozo. "Influence of geometric dimensions on the performance of a filtering hydrocyclone: An experimental and CFD study." *Brazilian Journal of Chemical Engineering* 26 (2009): 575–582.

Ottino, Julio M. and Devang V. Khakhar. "Mixing and segregation of granular materials." *Annual Review of Fluid Mechanics* 32 (2000): 55–91.

Pereira, Fabio A. R., Marcos A. S. Barrozo and Carlos H. Ataíde, C.H. "CFD predictions of drilling fluid velocity and pressure profiles in laminar helical flow." *Brazilian Journal of Chemical Engineering* 24 (2007): 587–595.

Perry, Robert H., and Don W. Green. *Chemical Engineers Handbook* (New York: McGraw-Hill, 1999).

Porter, S. J. "The design of rotary dryers and coolers." *Transactions of the Institution of Chemical Engineers* 41 (1963): 272–280.

Revol, D., Cedric L. E. Briens and Jean M. Chabagno. "The design of flights in rotary dryers." *Powder Technology* 121 (2001): 230–238.

Ristow, Gerald H. "Flow properties of granular materials in three-dimensional geometries". Habilitationsschrift, *Philipps-Universität Marburg* (1998): 63–92.

Rong, Wenjie, Baokuan Li, Yuqing Feng, Phil Schwarz, Peter Witt and Fengsheng Qi. "Numerical analysis of size-induced particle segregation in rotating drums based on Eulerian continuum approach." *Powder Technology* 376 (2020): 80–92.

Rose, Horace E., and R. M. E. Sullivan. *A Treatise on the Internal Mechanics of Ball, Tube and Rod Mills* (London: Constable, 35–68, 1957).

Santos, Dyrney A., Gustavo C. Alves, Claudio R. Duarte and Marcos A. S. Barrozo. "Disturbances in the hydrodynamic behavior of a spouted bed caused by an optical fiber probe: Experimental and CFD study." *Industrial & Engineering Chemistry Research* 51 (2012): 3801–3810.

Santos, Dyrney A., Marcos A. S. Barrozo, Claudio R. Duarte, Fabian Weigler and Jochen Mellmann. "Investigation of particle dynamics in a rotary drum by means of experiments and numerical simulations using DEM." *Advanced Powder Technology* 27 (2016b): 692–703.

Santos, Dyrney A., Fernando O. Dadalto, Rafael Scatena, Claudio R. Duarte and Marcos A. S. Barrozo. "A hydrodynamic analysis of a rotating drum operating in the rolling regime." *Chemical Engineering Research and Design* 94 (2015): 204–212.

Santos, Dyrney A., Claudio R. Duarte and Marcos A. S. Barrozo. "Segregation phenomenon in a rotary drum: Experimental study and CFD simulation." *Powder Technology* 294 (2016c): 1–10.

Santos, Dyrney A., Irineu J. Petri, Claudio R. Duarte and Marcos A. S. Barrozo. "Experimental and CFD study of the hydrodynamic behavior in a rotating drum." *Powder Technology* 250 (2013): 52–62.

Santos, Dyrney A., Rafael Scatena, Claudio R. Duarte and Marcos A. S. Barrozo. "Transition phenomenon investigation between different flow regimes in a rotary drum." *Brazilian Journal of Chemical Engineering* 33 (2016a): 491–501.

Santos, Kássia G., Valéria V. Murata and Marcos A. S. Barrozo. "Three-dimensional computational fluid dynamics modeling of spouted bed." *Canadian Journal of Chemical Engineering* 87 (2009): 2011–2019.

Schaeffer, David G. "Instability in the evolution equations describing incompressible granular flow." *Journal of Differential Equations* 66 (1987): 19–50.

Schofield, F. R., Glikin P. G. "Rotary coolers for granular fertilizer." *Chemical and Process Engineering Research* 40 (1962): 183.

Silva, Priscila B., Claudio R. Duarte and Marcos A. S. Barrozo. "Dehydration of acerola (Malpighia emarginata D.C.) residue in a new designed rotary dryer: Effect of process variables on main bioactive compounds." *Food and Bioproducts Processing* 98 (2016): 62–70.

Silverio, Beatriz C., Edu B. Arruda, Claudio R. Duarte and Marcos A. S. Barrozo. "A novel rotary dryer for drying fertilizer: Comparison of performance with conventional configurations." *Powder Technology* 270 (2015): 135–140.

Silverio, Beatriz C., Kássia G. Santos, Claudio R. Duarte and Marcos A. S. Barrozo. "Effect of the friction, elastic, and restitution coefficients on the fluid dynamics behavior of a rotary dryer operating with fertilizer." *Industrial & Engineering Chemistry Research* 53 (2014): 8920–8926.

Sunkara, Koteswara R., Fabian Herz, Eckehard Specht, Jochen Mellmann and Richard Erpelding. “Modeling the discharge characteristics of rectangular flights in a flighted rotary drum.” *Powder Technology* 234 (2013): 107–116.

Tada, Érika F.R., Andreas Bück, Evangelos Tsotsas and João C. Thoméo. “Mass transport in a partially filled horizontal drum: Modelling and experiments.” *Chemical Engineering Science* 214 (2020): 115448.

Walton, Otis R., and Robert L. Braun. “Simulation of rotary-drum and repose tests for frictional spheres and rigid sphere clusters.” *Proc. Joint DOE/NFS Workshop on Flow of Particulates and Fluids*, (1993): 1–18.

Wardjiman, Cherestella, Andrew Lee, and Martin Rhodes. “Behaviour of a curtain of particles falling through a horizontally-flowing gas stream.” *Powder Technology* 188 (2008): 110–118.

Watanabe, Hiroshi. “Critical rotation speed for ball-milling.” *Powder Technology* 104 (1999): 95–99.

Williams, John C. “The segregation of particulate materials: A review.” *Powder Technology* 15 (1976): 245–251.

Yang, Runyu Y., Aibing Yu, L. McElroy and Jie Bao. “Numerical simulation of particle dynamics in different flow regimes in a rotating drum.” *Powder Technology* 188 (2008): 170–177.

8 Prospects for the Development of the Industrial Process for Drying Nanoformulations

Eknath Kole and Sagar Pardeshi
Department of Pharmaceutical Technology, University Institute of Chemical Technology, Kavayitri Bahinabai Chaudhari North Maharashtra University Jalgaon, India

Arun Sadashiv Mujumdar
Department of Bioresource Engineering, Macdonald College, McGill University, Ste-Anne-de-Bellevue, Quebec, Canada

Jitendra Naik
Department of Pharmaceutical Technology, University Institute of Chemical Technology, Kavayitri Bahinabai Chaudhari North Maharashtra University Jalgaon, India

CONTENTS

DOI: 10.1201/9781003207108-8

8.1 INTRODUCTION

Drying is a method of removing volatile components thermally to produce a dry product. Drying is a critical process in the pharma, food, and cosmetic industries (Mujumdar, 2020). Drying processes employ removing liquid from a material through the vapor phase (Ishwarya, Anandharamakrishnan, and Stapley, 2015). Drying can aggravate nanoparticle aggregation upon evaporation, resulting in an initial reduction in the dissolution advantages provided by nanonization. Water-soluble or insoluble composite formers have been used in the drying process to reduce drying-cause agglomeration (Khor et al., 2017). It is still hard to develop adequate pharmaceutical unit operations for nanomaterials, including drying, granulation, as well as compaction, in a drying stage, which is essentially required to transform the liquid into solid forms. In the case of poor redispersibility, the drying process draws nanomaterials into close contact with one another, resulting in irreversible aggregation (Chung, Lee, and Lee, 2012). There are four basic stages in spray drying: liquid atomization, liquid mixing with the drying gas, liquid evaporation, and separation of the dehydrated particulate from the gas. The primary goal of this method of drying in pharmaceutical applications is to get dry particles with the required characteristics (R. K. Deshmukh and Naik, 2016). Particulate technology is a tactic for producing active substances in pure or matrix form under controlled settings. The primary goal of surface modification is to develop drug-containing particles with suitable physical and surface characteristics. It eliminates the need for excipient/carrier blends in formulations, thereby reducing cohesiveness. Surface energy, crystallinity, particle size, shape, surface roughness, and density are all affected by powder manufacturing techniques (Momin, Tucker, and Das, 2018). Particulate engineering is a relatively new field that covers all the aspects related to solid-state study, heat as well as mass transfer, life sciences, chemistry, pharmaceuticals, colloid and interface science, aerosol and powder science, and nanotech. The major function of the micronization and drying mechanisms was to accomplish appropriate size distribution and eliminate the majority of the solvent. Particulate engineering necessitates a more thorough understanding of particulate formation mechanisms (Vehring, 2008). Numerous challenges in pharmaceuticals can be solved by nanotechnology, especially the challenge of drugs' low aqueous solubility (Wais et al., 2016). The latest developments in nanoscience, particulate engineering, and material science have unlocked the way to improve the efficacy of aerosolized therapy in respiratory conditions. To generate respirable particles, various particle engineering approaches are useful for example, spray drying, freeze drying, ultrasound-assisted antisolvent crystallization, and high-volume methods of production (Schuck, 2009; Mehta et al., 2018). The powder form is versatile enough to allow for advanced formulation. Furthermore, the generation of powders may meet the needs of the pharmaceutical, food, and cosmetic industries (Shishir and Chen, 2017). Spray drying is a process for obtaining particle forms in dry powder on an industrial scale. Spray drying would be a concise one-step procedure to producing powder particles on a big scale, converting a liquid feed into dry particles by atomizing the feed into a hot, dry chamber (Liu, Chen, and Selomulya, 2015; Singh and Van den Mooter, 2016). All drying methods put probiotics under stress, limiting their survival (Moayyedi et al., 2018). Spray drying is a novel, easy, rapid, and repeatable drying method to obtain nanoparticles (Zahariev, Pilicheva, and Simeonova, 2019). APIs are frequently preserved by encapsulation using drying methods. Nano-spray drying, a novel drying technology, is being established to achieve these properties. Due to the varying features of particulates in different formulations such as suspension, fat, and oil embedded in dispersion or fine emulsion, every preparation may necessitate distinct processibility to produce free-flowing dry powders with intended product attributes (Jacobs, 2014). As a result, spray drying is broadly used for the development of drug formulation for drug delivery via the respiratory, mucous, skin, and oral routes of administration (P. A. Ferreira et al., 2020). Among the numerous microencapsulation techniques, spray drying is the most common useful process to stabilizing active molecules by protecting them in different carriers, as well as enabling continuous production of dry materials (Sarabandi, Gharehbeglou, and Jafari, 2020; de Souza et al., 2020). Presently, commercial drug nanoformulation manufacturing is classified into two classes i.e. 1) top-own approaches, and 2). bottom-up strategies. First strategies, such as wet media milling and high-pressure homogenization, convert bigger crystalline particles into nanoparticles. Physical attrition or grinding is used to minimize the size of the particles in top-down methods. Bottom-up strategies lead to the formation of solid nanoparticles either by precipitation or by solvent evaporation, in which particles are developed from a molecular state, such as precipitation by liquid solvent-antisolvent addition, supercritical fluid precipitation, spray-freezing, and nano-spray drying (Hu et al., 2018).

8.1.1 Spray Drying and Nano-Spray Drying

8.1.1.1 Spray Drying

The spary method is widely used to dry temperature-sensitive foods, medical products, as well as other substances (Waghulde and Naik, 2017; Waghulde et al., 2019;

Khairnar et al., 2020; Verma, Mujumdar, and Naik, 2020). Spray drying is a single-step process where a fluid sample has been atomized inside a gas that is an extremely hot stream to produce a dry product instantly (R. Deshmukh, Wagh, and Naik, 2016; P. S. Wagh and Naik, 2016; Khairnar et al., 2020). Spray-drying encapsulation has received a lot more recognition from scientists all over the world than most other microencapsulation methods for its significant advantages over other micro-encapsulation methodologies (R. K. Deshmukh and Naik, 2014; P. Wagh, Mujumdar, and Naik, 2019). Nanoparticles are usually produced using a traditional approach like solvent evaporation, spray drying, ultrafine milling, and high-speed homogenization (Patil, Khairnar, and Naik, 2015; Shrimal et al., 2019). Among others, spray drying may be a simple unit operation that changes a liquid feed onto an intended engineered powder product (P. S. Wagh and Naik, 2016; R. Deshmukh, Mujumdar, and Naik, 2018). In consequence, spray drying has been widely used to dry pharmaceuticals with low moisture content and specific particle sizes for targeted therapy of the lungs. It is imperative to optimize process parameters to avoid the issue of low yields, high moisture contents, or sticking of products due to spray drying. The efficiency of the spray-drying method is heavily reliant on variables such as spraying nozzle diameter, inlet and outlet temperature, and aspirator speed (Pardeshi et al., 2020).

8.1.1.2 Nano-Spray Drying

Buchi technology was first introduced as nano-spray drying equipment in 2009. Since its inception, the dryer has been used for scientific research, with the main emphasis on the development of pharmaceutical applications as well as the food industry. The drying principle is comparable to that of conventional spray drying (Piñón-Balderrama et al., 2020). The nano-spray drying method, as opposed to the spray-drying technique, is appropriate for drying temperature-sensitive products because it relies on innovative spray mesh technology (Zahariev, Pilicheva, and Simeonova, 2019). On a laboratory level, nano-spray drying has a broad range of pharmaceutical applications, including improved absorption of drugs with low solubility through surface modification, and entrapment before nanoparticles and nanoemulsion, and also nanosuspension via polymeric biomaterial as prolonged drug release. The drying process is delicate, which helps with the significant stability and potency of thermally sensitive materials. Drug-loaded particles prepared by nano-spray drying are given via respiratory, oral, injectable, topically, ocular, IP, and IVES routes, demonstrating every technology's flexibility. The primary applications include respiratory drug delivery, nanotherapeutics, entrapment of nanoemulsion with hydrophobic drugs, and the preparation of nanocrystals for improved absorption. The purpose of nano-spray drying allows the entrapment of active compounds in a polymeric matrix, which provides improved environmental protection, stability, handling, storage, and controlled drug delivery (Crpagaus Arpagaus, 2018). The nano-spray drying technique is effective for forming sub-micron polymer powders containing bio-active ingredients. Considerable research on nano-spray drying taking place, in pharmaceutical fields along with material technology and the food industry (Cordin Arpagaus, 2019a). Nano-spray drying was exploited to obtain ultrafine powders from colloidal SLNs (Xue et al., 2018). In numerous delivery systems, the conventional dosage form can be transformed into a dry powdered dosage. As nanoencapsulation strategies progressed, the nano-spray drying technique was developed to improve the manufacturing or administration of drugs containing solid colloidal particles of sub-micron size. Encapsulating a drug in a polymeric matrix improves environmental protection, such as resistance to oxidation, light, and temperature, as well as drug stability, control, storage, and control release of drug (Cordin Arpagaus, 2018).

8.1.2 Differences and Developmental Gaps for Nano-Spray Drying

Various attributes related to a traditional spray dryer and a nano-spray dryer for laboratory-scale are given in Table 8.1. A non-spray dryer technique requires small sample amounts to obtain fine particles as well as to maxims product yield. This technique is presently suitable for laboratory scales.

8.1.3 Nano-Spray Drying Setup

A small-scale spray dryer is a spray drying method that can produce sub-micron particles with solution, nanoemulsion, and nanosuspension. A spray-drying

TABLE 8.1
Key Differences Between a Traditional Spray Dryer and a Nano-Spray Dryer on a Laboratory Scale [Reproduced from (Cordin Arpagaus et al., 2017) with Kind Permission of the Copyright Holder, Elsevier, Amsterdam]

Attributes	Traditional Spray Dryer	Nano-Spray Dryer
Important goals and advantages	Traditional spray drying using a tried-and-true method	Required small sample amounts, obtain fine particles, and maximize yield
Key Spray techniques	Two fluid or ultrasonic nozzles	A vibrating mesh atomizer powered by piezoelectric energy produces a fine spray of particles.
Drying gas flow regime	Turbulence	Laminar
Particle separation technology	Cyclone	Electrostatic particle collector
Drying temperature	Temperature ranges of up to 220°C	Temperature ranges of up to 120°C
Sample viscosity	Not more than 300 cycles per second	Not more than 10 cycles per second
Droplets size	5 to 100 microns, wider	5 to 15 microns, small
Dried particles size	For two-fluid nozzle 2–25 micron (two-fluid nozzle), and ultrasonic nozzle i.e., 10 to 60 micron	0.2 to 5 microns
Scale-up capability	Simplified scale-up from lab to pilot (kg scale) and commercial scale (ton scale)	Presently suitable to lab scale (gm scale)

Source: Data summarized from (Büchi Labortechnik AG, 2017).

mechanism is based on vibrating mesh (spray technology), along with an extremely effective electrostatic precipitator to obtain nanoparticles as its technical advances. A nano-spray dryer with its operational principle is illustrated in Figure 8.1. There are four fundamental steps involved in nano-spray drying as below, i.e., a) heating of the gas, b) droplet generation, c) droplet drying, and d) particle collection. Conventional spray dryers are limited in their yields, reaching at most 50%–70%. Additionally, to perform a feasibility test, a minimum amount of liquid sample of 30 mL must be present. However, a very few milligrams of the sample that can be used in a nano-spray dryer is enough.

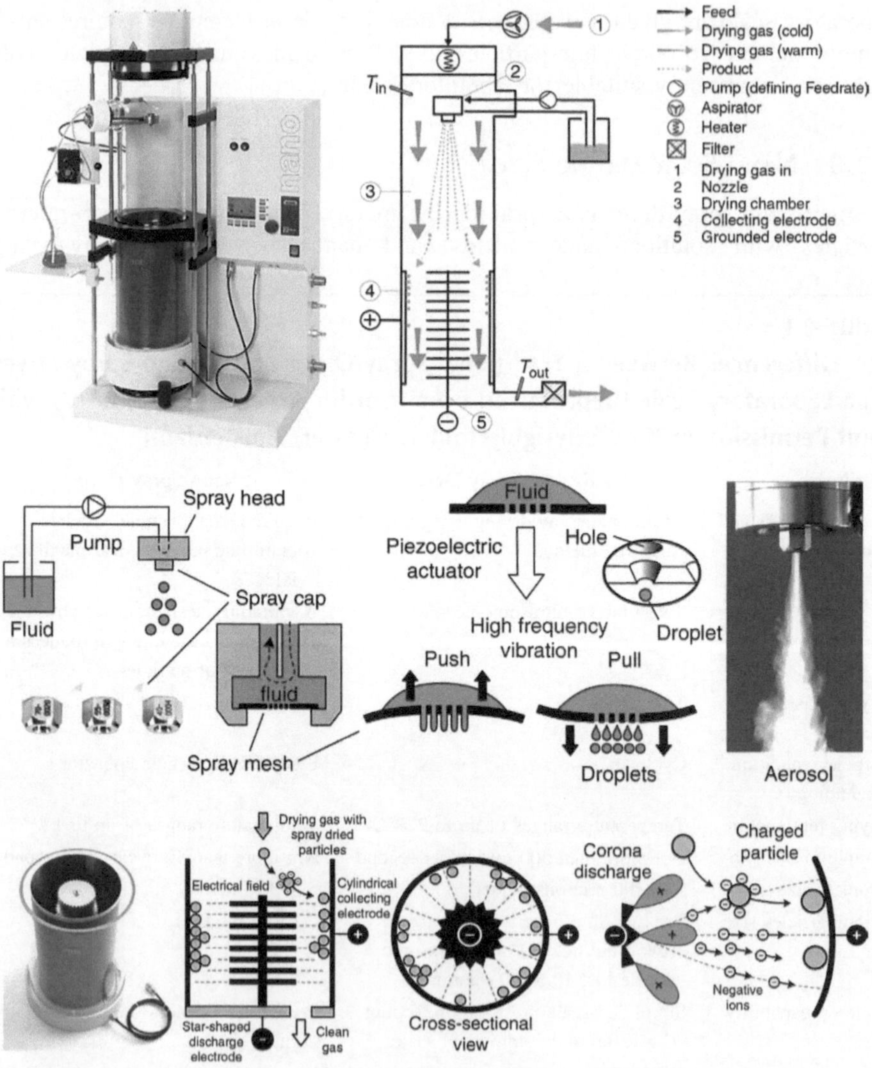

FIGURE 8.1 The nano-spray dryer with its operational principle. [Reproduced from (Cordin Arpagaus et al., 2017) with kind permission of the copyright holder, Elsevier, Amsterdam].

8.2 PROCESS CAPABILITIES AND FUNDAMENTALS OF NANOFORMULATION DRYING

8.2.1 Process Variables and Formulation Variables

For the optimization of nanoformulation, various factors can affect the characteristics of nanoformulation. There is a need for optimization of formulation and process parameters of nanoformulation. Various variables are discussed in this section. Process variables like formulation and processing variables and their outcomes are as follows.

8.2.1.1 Influences of Process Parameters

In nano-spray drying, a connection or dependency on process variables as well as formulation parameters determine the characteristics of the generated powder. Figure 8.2 indicates primary process formulation factors for the nano-spray dryer and the outputs.

Some regulations for optimizing the nanocapsule formation process were recognized and are listed in the subsequent sections. Some preliminary process conditions for nano-spray drying are listed (Naar et al., 2017). Usually, process variables are optimized through trial and error (Harsha et al., 2017). The impacts on the size of particles, yield, as well as output produced per second are critical (Cordin Arpagaus et al., 2018).

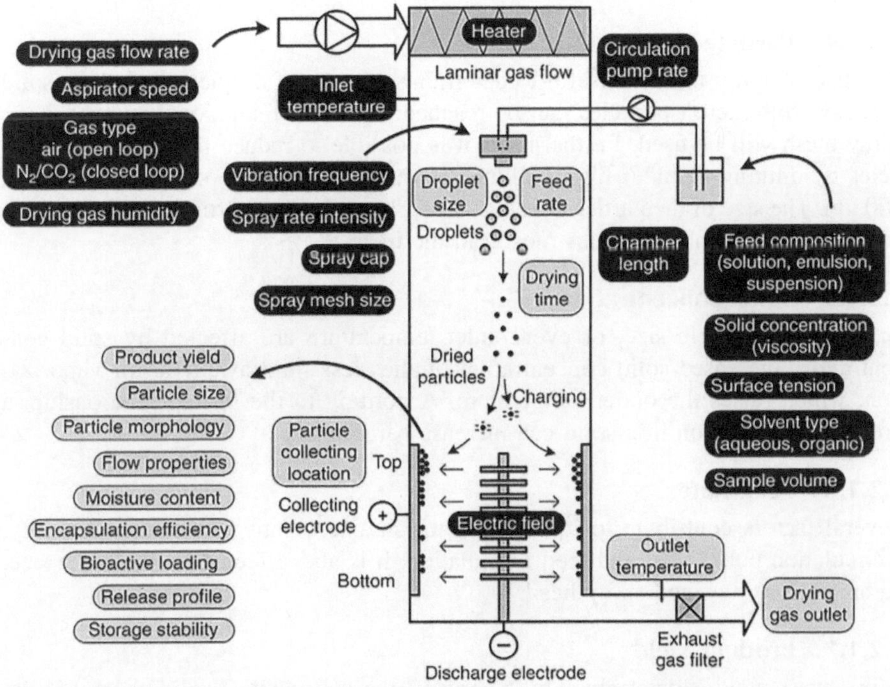

FIGURE 8.2 Primary process formulation factors for the nano-spray dryer and the outputs. [Reproduced from (Cordin Arpagaus et al., 2017) with kind permission of the copyright holder, Elsevier, Amsterdam].

Particulate morphology, drug loading, controlled drug release pattern, entrapment efficiency, and sustainability are some of the output variables of interest.

8.2.1.2 Drying Gas Flow Rate, Humidity, and Temperature

The flow rate of drying gas temperature, along with liquid feed rate regulate the outlet gas temperature. The higher the inlet temperature, the lower the relative humidity of drying gas, which gives greater yield, and not as much viscous powder at the output. Whenever the spray rate intensity is reduced from 100% to 25%, the particle size is reduced to approx 7.7–5.5 m, and also the timescale has risen to 1.8–2.1 m. By adjusting the aspirator speed, the drying gas flow rate can be controlled to a range of 80–160 L/min. A significantly greater humidity level within drying gas can reduce the capability to absorb vapor, which tends to result in a marginally higher outlet temperature.

8.2.1.3 Droplet Size

To regulate the size of a droplet, characteristics such as atomized fluid, which includes viscosity and surface tension, should be taken into account. A reduced flow rate can be attributed to higher viscosity. Following that, a smaller fraction of fluid volume may be supplied during mesh vibration, resulting in the formation of tiny droplets. Organic solvents typically produce tiny droplets owing to low surface tension, flowability, as well as density.

8.2.1.4 Particle Size

The size of spray mesh is useful to determine the size of droplets as well as solid particles. Submicron particles can be reached, by diluted solutions, and a 4.0 μm spray mesh will be used. Furthermore, it is possible to reduce the size of the particles by diluting them. A highly diluted solution can produce small particles i.e., 100 nm. The size of the particles results in nearly equivalent particle diameter or the concentration of solid particles mostly in the feed.

8.2.1.5 Solid Concentration

A feed rate, particle size, or even outlet temperature are affected by solid concentration. Increased solid concentration implies less fluid available for vaporization, which raises the outlet temperature. According to the former relationship, a greater concentration in a solid can increase particle size.

8.2.1.6 Feed Rate

Several factors contribute to it, including mesh size, relative spray rate intensity, recirculation pump rate, and feed formulation. It is also affected by the mesh size, parameter settings, and spray head.

8.2.1.7 Product Yield

A product's yield is determined by measuring the intermediate solid content in the feed and the weight of the obtained powder after nano-spray drying. Under optimal conditions, a 100 mg sample can achieve a yield of 76%–96% for various drugs.

Using the nano-spray dryer makes it possible to experiment with valuable biological materials since it can process limited amounts of samples.

8.2.1.8 Organic Solvent Instead of Water

In nano-spray drying, several organic solvents are used, each chosen according to its solubility and ability to incorporate wall materials. Dexamethasone dissolves well in an acetone-water mixture because the acetone has a low viscosity, it can be fed much more rapidly through vibrating mesh, reducing processing time. Dichloromethane, a solvent, is broadly used for drug encapsulation because it has a high PLGA solubility and can be processed at a reduced inlet temperature, which can overcome problems such as sticky output. As a result, in ambient conditions, DCM can evaporate and particles can be formed, making it an ideal process for heat-sensitive compounds.

8.2.1.9 Encapsulation Efficiency and Active Compound Loading

A nano-spray drying method for nanoencapsulation involves choosing the wall material based on microencapsulation factors. Numerous parameters for nanocapsules include entrapment efficiency and active compound loading. More than 99% of drugs can be encapsulated in nanocapsules with adjustable drug loading.

8.2.1.10 Controlled Release of Active Compounds

The polymer used, its molecular weight, and particle size do have an impact on the release mechanism. Because of the distinction in exposed surface area, smaller particles will release the active substance faster than larger particles, as demonstrated by Beck-Broichsitter et al. (2015). Drugs can be released for up to six weeks from polymers with small molecular weight, such as PLGA. Typically, the release pattern shows two phases: a burst phase, when 10%–30% of the active ingredient is released because of free bounding on the surface, and a prolonged phase in the second stage (Cordin Arpagaus et al., 2017). The process variables and their effects on output parameters i.e., various process parameters like outlet temperature, droplet size, feed rate, moisture content, yield, and stability.

8.2.2 Particle Morphology and Surface Characteristics

By studying the morphology and crystallization states of drugs, we can assess their morphological or structural changes during nanosizing. During the preparation of nanosuspensions, amorphous drug nanoparticles are highly likely to be formed. Therefore, it is vital to analyze their numbers. Amorphous forms of the chemical are inherently unstable and are prone to crystallization when stored. When preparing or formulating nanosuspensions, such a transition over storage must be considered. X-ray diffraction analysis and scanning electron microscopy have been widely used to determine the crystalline phase and particle morphology of drugs (Zhang et al., 2017).

8.2.2.1 Particle Morphologies

With the help of scanning electron microscopy, one can observe the particle morphology, i.e., particles size, shape, and surface structure of nano-spray dried powders. Fluorescence optical microscopy can be used to affirm appropriate encapsulation.

Particles prepared by nano-spray drying exhibit various morphology. The shape of nano-spray-dried particles is determined by feed properties (for example, material type, solid concentration, solvent, and surfactant) and drying situations (e.g., temperatures). The shapes of the various particles developed with the help of nano-spray drying are depicted in Figure 8.3. One of the most stable forms of a droplet is the

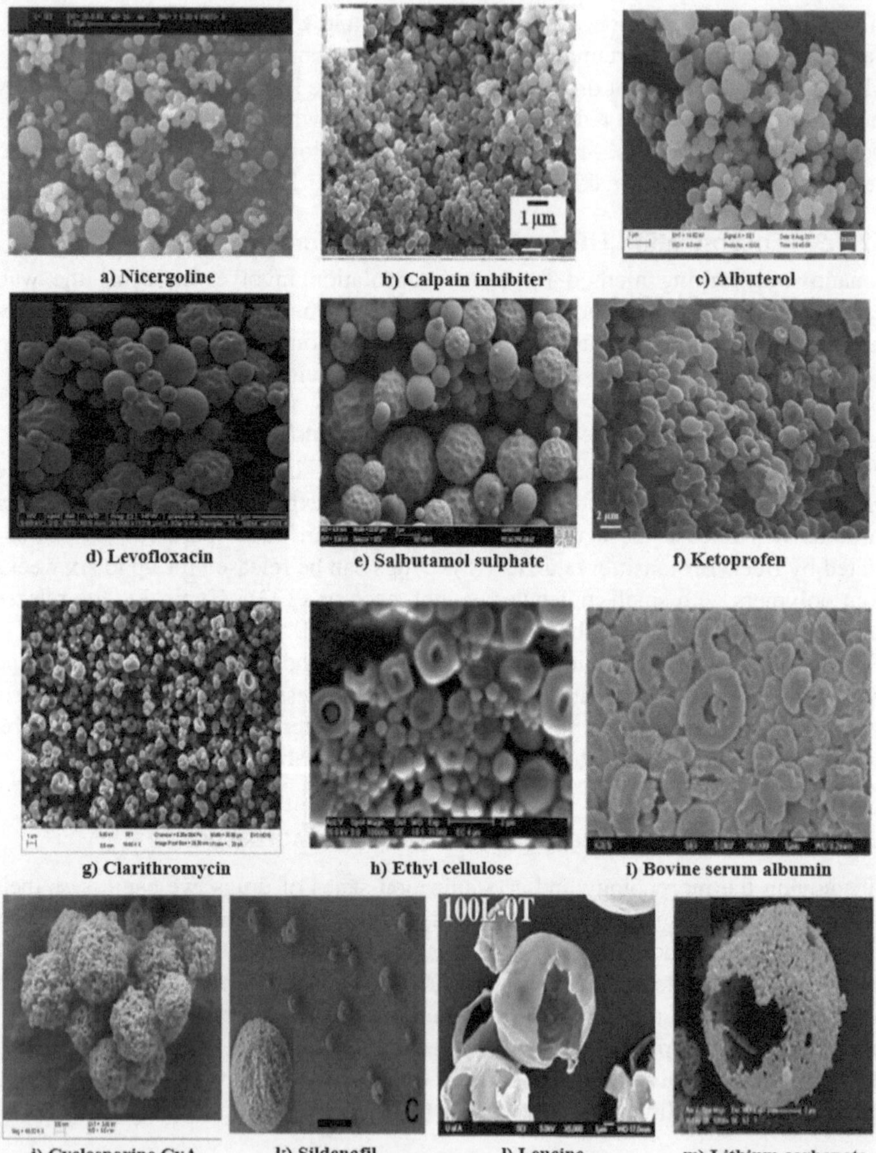

FIGURE 8.3 Spherical particles (a, b, c), shriveled/wrinkled particles (d, e, f, g), doughnut-type particles (h, i), composite-type particles (j, k), hollow particles (l, m) prepared by nano-spray drying. [Reproduced from (Cordin Arpagaus et al., 2017) with kind permission of the copyright holder, Elsevier, Amsterdam].

spherical shape of nano-spray-dried particles. As a result, compact spherical particles are supposed to be created during nano-spray drying. Surfactants can be used to improve the spherical geometry of particles. Surfactant-free particles have shrunk or wrinkled edges. Large particles, in general, have shrunk surfaces. Doughnut-like particles form as a result of hydrodynamic effects and structural instability. This may take place due to high viscosity, which may impact the morphological characteristics of nano-spray-dried particles. As a result, because the doughnut shape has a larger surface area than the spherical shape, it is not prepared in encapsulation. Before actually nano-spray drying, nanoparticle suspensions form leads to the formation of composite particles; composite particles include flocculated nanoparticles with a high surface area. Furthermore, nanoparticles can retain their integrity, characteristics, redispersibility, increased solubility, and oriented capacity. Due to high temperatures and rapid water loss, hollow particles with thin shells form. The creation of hollow particles can be aided using floor-active agents such as leucine and the incorporation of organic solvents into the formulation (Cordin Arpagaus et al., 2017). Spherical and doughnut shape ethyl cellulose and spherical poly (lactic-co-glycolic acid) carrier particles were generated by Nano-Spray Dryer B-90. Nano-spray-dried EC particles showed spherical and partially doughnut-like morphology, while PLGA particles have only a spherical shape using a hole size 7.0 μm and 5.5 μm spray caps (Dahili and Feczkó, 2015).

8.2.2.2 Surface Characteristics

Inadequate characterization has an impact on many areas of nanomaterial research, including repeatability of outcomes, the method of scaling up to fulfill manufacturing needs, and the assessment of human health and environmental risks. Surface analysis methods with multiple sources are underutilized for the characterization of nanostructured materials. The conventional method of surface analysis generates a variety of useful data about nanostructures from what has been obtained and collected, which is what many analysts anticipate (Baer et al., 2013). Meeus et al. (2013) prepared spray-dried PLGA/PVP microspheres. This observation considered the effect of heat and humidity on the surface characteristic. The said point could provide perspective into expected release behavior as well as a stable state of formulation based mostly on polymeric metrics. As a result, a drug matrix with desirable and tunable properties in terms of state physical and chemical stability or release of drug within the development state. Glass transition temperatures and miscibility behavior were studied using DSC. AFM has been utilized to investigate the nanoscale topography and phase behavior of samples. TOF-SIMS and XPS were used to assess and measure the chemical properties of the surface. Moisture and temperature exposure influenced the surface characteristics of spray-dried PVP/PLGA nanoparticles. Moreover, a potent combination of cutting-edge technology enables nanoscale surface characterization (Meeus et al., 2013).

8.2.3 Advantages and Limitations of Nano-Spray Drying

There are several advantages of spray drying. Its delicate operation can handle a wide range of compounds, such as thermosensitive substances like biologics,

TABLE 8.2
Advantages and Disadvantages of the Spray-Drying Technique [Reproduced from (Tania Marante, Ines Duarte, Macedo, and Fonte, 2020)

Advantages	Disadvantages
Control over particle morphology is ideal.	Produce a low yield
Powder characteristics are well-governed.	Yields mainly amorphous powders
Adaptability.	Thermosensitive substances should not be used.
Appropriate for the solution, suspension, or emulsion-based system.	Physical instability
Handy for molecules that are either water-soluble as well as insoluble.	

pharmaceuticals, and food nutrients (Büchi Labortechnik AG, 2017; Abdel-Mageed et al., 2020). Spray drying is a desirable process for obtaining particles because it is a fairly simple, quick, constant, reproducible, scalable, and cost-effective method (Sosnik and Seremeta, 2015; Cordin Arpagaus, 2019c). The above technique is a one-step process that is suitable for heat-labile substances and allows for control over particle morphology. This process has the potential to provide quick evaporation as a cooling effect on preparations (Haggag and Faheem, 2015). The small droplet formed throughout atomization is exposed to high temperatures for a short period during the drying phase, making it more suitable for thermolabile substances (Sosnik and Seremeta, 2015). This method is used to improve drug bioavailability as well as drug life span (long-term stability) (Fonte et al., 2015, 2016; Fonte, Reis, and Sarmento, 2016). The spray-drying technique can be used to produce highly effective drug-loaded nanoparticles. It is best suited for use with DPI (dry powder inhalers) to deliver drugs directly to the lungs (J. Broadhead, 2013; Haggag and Faheem, 2015; Sosnik and Seremeta, 2015; Tania Marante, Ines Duarte, Macedo, and Fonte, 2020). Spray drying has several limitations, including the possibility of causing thermal damage to a certain molecule, such as proteins, that can cause instability or a modify in one chemical structure. Due to shear stress, the said drying condition can cause modification and denaturation of protein, which might also influence stability. Product yield is dependent on work scale, to lower scale setups obtaining low output due to product loss inside the drying chamber wall. Too much sample was needed, and a faster rate of solvent evaporation. Moreover, higher pumping power has been required for pumping solvent feed (Haggag and Faheem, 2015; Sosnik and Seremeta, 2015; Tania Marante, Ines Duarte, Macedo, and Fonte, 2020). Various advantages and disadvantages of spray-drying techniques are summarized in Table 8.2.

8.3 STABILITY OF FORMULATIONS DURING NANO-SPRAY DRYING

Submicron drug particles have significantly greater physical and chemical stability when compared to other existing formulations, as well as superior solubility and bioavailability. One of the most extensively researched liposome formulations is SLN and NLC, which is an enormous ability for generic delivery systems and excellent drug and nutrient stability. Usually lipid particles layered with pectin and gum arabic have been investigated during spray drying for the preparation of stable SLN and NLC. The zeta potential of all prepared lipid particles ranged from −50 to −80 mV, showing strong repulsion from among particles in this dispersion, indicating excellent colloidal stability. Polysaccharides were effectively using it as a coating material by the research scientist to prepare stable SLN and NLC (Wang et al., 2016). Maged et al., (2017) used a nano-spray dryer to develop econazole nitrate nanoparticles in weight ratios of 1:1, 1:2, and 1:3 of the drug to hydroxypropyl-cyclodextrin and stabilizer. The formulation was sprayed through a 7.0 μm nozzle with inlet and outlet temperatures of 95°C and 45°C, respectively. At room temperature, the drug particles were uniformly distributed in an isotonic buffer solution and tested for stability. When compared to other preparations, the econazole nitrate nanosuspension with a 1:1 weight ratio had the greatest stability during storage at room temperature. Because of its good balance for the stability of drug-loaded nanoparticles, the optimum weight ratio of 1:1, drug to hydroxypropyl-cyclodextrin, was selected as the best weight ratio (Maged, Mahmoud, and Ghorab, 2017). Before drying, maltodextrin was added to the fish oil emulsion and hydroxypropyl methylcellulose solutions, which improved chemical stability. Moreover, increasing the emulsion and particles size of powder up to 10–150 mm often can decrease particle surface area and, as a result, enhance storage stability. The scavenging activity of free rutin was concentration-dependent, with higher rutin concentrations leading to increased antioxidant properties. Slight variations in the antioxidant properties of free rutin were witnessed throughout the assay (72 hours), suggesting the molecule's stability in the medium (Pedrozo et al., 2020). Particle engineering ensures adequate stability throughout storage (Mukhopadhyay, 2018). The manufacturing of stable spray-dried lipid NPs (LNP) would enhance shelf-life stability. LNP's stability properties may be advantageous for use in a variety of industrial biocatalytic and biotherapeutic fields (Abdel-Mageed et al., 2020). The drying process using nano-spray drying is gentle and makes a significant contribution to the stability and activity of thermolabile substances like peptides (Crpagaus Arpagaus, 2018; Cordin Arpagaus, 2019b). Spray-drying encapsulation (SDE) is useful since it converts liquid feeds into dry powder with greater stability. Another alternative is to include natural antioxidants in the feed formulation, which can prolong the shelf life of produced entrapped oil powders (Assadpour and Jafari, 2019). Entrapment methods, like emulsification, are typically used in conjunction with drying methods to avoid biological or chemical deterioration, thereby improving suspension stability and extending the shelf-life of food products (Nunes et al., 2020).

8.4 APPLICATION OF SPRAY AND NANO-SPRAY DRIED NANOFORMULATIONS

8.4.1 Solubility Enhancement of Poorly Soluble Drugs

Nano-spray drying has been used effectively in various drug delivery applications, including raising the solubility and bioavailability of hydrophobic drugs through nanonization and structural modification and encapsulating nanoparticles in biocompatible polymers for prolonged drug release (Cordin Arpagaus, 2019b). Approximately 60% of all synthesized active compounds are poorly water-soluble. As a result, it seems that the nanocrystal technique will continue to flourish as a valuable tool in pharmaceutics to increase the solubility of drugs, oral absorption, or bioavailability (Gülsün, Gürsoy, and Öner, 2009). Nanoparticles developed over the last two decades as potential candidates for enhancing the water solubility of drugs. The spray-dried glipizide nanosuspension was successfully prepared to improve glipizide solubility and dissolution (Elham et al., 2015). Spray drying techniques were used to produce a dry powder from an itraconazole nanosuspension, then compressed into tablets. In conclusion, the nanosuspension method can enhance the oral absorption and bioavailability of tablet dosage forms containing BCS Class II drugs (Sun et al., 2015). TMS is used to lower blood pressure. The researchers aimed microenvironmental pH to enhance solubilization performance to improve TMS solubility (Miller, Ellenberger, and Gil, 2016). Nicergoline NP has been spray dried using a nano-spray dryer and water: ethanol as a solvent system to generate nicergoline NPs in an amorphous form to improve solubilization (Maged, Mahmoud, and Ghorab, 2017). Darunavir and Felodipine amorphous solid dispersion were developed to improve solubility using spray drying (Salama, 2020).

8.4.2 Encapsulation of Pharmaceuticals and Biopharmaceuticals

Nano-spray drying is gaining popularity in a wide range of therapeutic applications. Trehalose as a shelf-life stabilizer, as a dispersant, the *l*-leucine amino acid is used. Hypromellose is a viscoelastic polymer, and chitosan or PLGA as biodegradable materials are examples of typical nano-spray-dried pharmaceutical additives. Due to the apparent advantages of nanonization and structural change, numerous studies show that nano-spray drying is also best suited for transforming pure drugs into dry nanoparticles. Ethanol, methanol, and acetone are common solvents used to solubilize poorly soluble drugs in water. As encapsulation wall materials, a variety of polymers were used. Chitosan, gelatin, methylcellulose, lactose, a starch derivative, PVP, as well as polyvinyl alcohol are all examples of water-soluble polymers. Furthermore, water-insoluble polymers are widely employed as encapsulating agents. In another research, terbutaline sulfate and antibiotics such as ciprofloxacin and azithromycin show that this improved excipient-growth approach is feasible (Cordin Arpagaus et al., 2017; Büchi Labortechnik AG, 2017).

8.4.3 Encapsulation of Nutraceuticals

A nano-spray dryer has primarily been used in pharmaceutical and material science applications (Li et al., 2010). Nanoemulsion of vitamin E acetate was prepared, in which gum and starch were used to incorporate tocopherol. Tocopherol was selected as the oil phase model. From this, we can conclude that the entrapment of nanoemulsion by nano-spray drying maintained the stability of the prepared nanoemulsion. Hence, it is considered novel as well as a promising method of delivering nano-emulsions (Li et al., 2011). In addition, Perez-Masia et al. recently presented an application of vitamin B-9 entrapment with whey protein and resistant starch. Folate is a polar compound and found in various foods. They were able to create spherical submicron capsules (Pérez-Masiá et al., 2015). Research should focus to develop polymeric encapsulants with high entrapment characteristics.

8.5 CURRENT CHALLENGES AND FUTURE PERSPECTIVES

Some liquids are excessively thick to pass through the mesh system. Droplet generation may be irregular or even cease when mesh holes are restricted by the suspension. In this case, the processing time will slowly increase, and the product may occasionally accumulate on the vibrating mesh, reducing production. Large-scale nano-spray drying equipment has yet to be established and commercialized. More research is required to determine the scalability of the processing parameters. Electrostatic particle collectors, like mechanical rapping devices, are equipped with an automated cleaning system to continually abolish particles silt on collector electrodes before they reach saturation (Cordin Arpagaus et al., 2017). There are phenomenal developments in the application of nanotechnology in pharma, cosmetic, and food research. Nanotechnology rules and regulations can conquer safety-related difficulties associated with drying methods, and it has the potential to rule the entire pharmaceutical industry and the cosmetic and nutraceutical domains. It is expected that nanotechnology will become an advanced technology by 2050, solving most industrial and societal problems due to its ability to find cordial solutions at both the micro and macro levels. In the future, further research should explore how nano-spray drying might be industrialized. It's becoming increasingly important to scale up. Nanomedicine is primarily developed for respiratory drug delivery, nanotherapeutics, encapsulating poorly water-soluble active ingredients in nanoemulsion, and formulating nanocrystals to enhance bioavailability.

8.6 CONCLUDING REMARKS

A nano-spray drying methodology employs advanced spray mesh technology in comparison to the traditional spray-drying method. This technique can lead to different particle morphologies and particle characteristics. In the field of pharma, nano-spray drying is commonly used to enhance the bioavailability of drugs through encapsulation in polymeric nanoparticles. This technology helps to overcome problems associated with the low bioavailability of drugs. The drug particles prepared through spray drying and nano-spray drying are administered in a variety of ways.

ACKNOWLEDGEMENT

One of the authors is grateful to the University Grants Commission (UGC), New Delhi, for providing the financial support, in terms of UGC-BSR Mid-Career Award.

REFERENCES

Abdel-Mageed, Heidi M., Shahinaze A. Fouad, Mahmoud H. Teaima, Rasha A. Radwan, Saleh A. Mohamed, and Nermeen Z. AbuelEzz. 2020. "Engineering Lipase Enzyme Nano-Powder Using Nano Spray Dryer BÜCHI B-90: Experimental and Factorial Design Approach for a Stable Biocatalyst Production." *Journal of Pharmaceutical Innovation* 16: 759–771. 10.1007/s12247-020-09515-4

Arpagaus, Cordin. 2018. "A Short Review on Nano Spray Drying of Pharmaceuticals." *J Nanomed Nanoscience* 3 (4): 149. 10.29011/2577-1477.100049

Arpagaus, Cordin. 2019a. "Nano Spray Drying of Bioactive Food Ingredients." *Proceedings of Eurodrying'2019*, July: 10–12.

Arpagaus, Cordin. 2019b. "Nano Spray Drying of Pharmaceuticals," no. September: 11–14. 10.4995/ids2018.2018.7356

Arpagaus, Cordin. 2019c. "PLA/PLGA Nanoparticles Prepared by Nano Spray Drying." *Journal of Pharmaceutical Investigation* 49 (4): 405–426. 10.1007/s40005-019-00441-3

Arpagaus, Cordin, Philipp John, Andreas Collenberg, and David Rütti. 2017. *Nanocapsules Formation by Nano Spray Drying. Nanoencapsulation Technologies for the Food and Nutraceutical Industries*. 10.1016/B978-0-12-809436-5.00010-0

Arpagaus, Cordin, Andreas Collenberg, David Rütti, Elham Assadpour, and Seid Mahdi Jafari. 2018. "Nano Spray Drying for Encapsulation of Pharmaceuticals." *International Journal of Pharmaceutics* 546 (1–2): 194–214. 10.1016/j.ijpharm.2018.05.037

Arpagaus, Crpagaus. 2018. "Pharmaceutical Particle Engineering via Nano Spray Drying - Process Parameters and Application Examples on the Laboratory-Scale." *International Journal of Medical Nano Research* 5 (1): 5–26. 10.23937/2378-3664.1410026

Assadpour, Elham, and Seid Mahdi Jafari. 2019. "Advances in Spray-Drying Encapsulation of Food Bioactive Ingredients: From Microcapsules to Nanocapsules." *Annual Review of Food Science and Technology* 10: 103–131. 10.1146/annurev-food-032818-121641

Baer, Donald R., Mark H. Engelhard, Grant E. Johnson, Julia Laskin, Jinfeng Lai, Karl Mueller, Prabhakaran Munusamy, et al. 2013. "Surface Characterization of Nanomaterials and Nanoparticles: Important Needs and Challenging Opportunities." *Journal of Vacuum Science & Technology A: Vacuum, Surfaces, and Films* 31 (5): 050820. 10.1116/1.4818423

Broadhead, J., S. K. Edmond Rouan, and C. T. Rhodes. 2013. "The Spray Drying of Pharmaceuticals." *Journal of Chemical Information and Modeling* 53 (9): 1689–1699.

Büchi Labortechnik A. G. 2017. "Nano Spray Drying Booklet – Theory & Applications," 1–40.

Chung, Nae Oh, Min Kyung Lee, and Jonghwi Lee. 2012. "Mechanism of Freeze-Drying Drug Nanosuspensions." *International Journal of Pharmaceutics* 437 (1–2): 42–50. 10.1016/j.ijpharm.2012.07.068

Dahili, Laura Amina, and Tivadar Feczkó. 2015. "Cross-Linking of Horseradish Peroxidase Enzyme to Fine Particles Generated by Nano-Spray Dryer B-90." *Periodica Polytechnica Chemical Engineering* 59: 209–214. 10.3311/PPch.7590

Deshmukh, Rameshwar, Pankaj Wagh, and Jitendra Naik. 2016. "Solvent Evaporation and Spray Drying Technique for Micro- and Nanospheres/Particles Preparation: A Review." *Drying Technology* 34 (15): 1758–1772. 10.1080/07373937.2016.1232271

Deshmukh, Rameshwar, Arun Mujumdar, and Jitendra Naik. 2018. "Production of Aceclofenac-Loaded Sustained Release Micro/Nanoparticles Using Pressure Homogenization and Spray Drying." *Drying Technology* 36 (4): 459–467. 10.1080/07373937.2017.1341418

Deshmukh, Rameshwar K., and Jitendra B. Naik. 2014. "Study of Formulation Variables Influencing Polymeric Microparticles by Experimental Design." *ADMET and DMPK* 2 (1): 63–70. 10.5599/admet.2.1.29

Deshmukh, Rameshwar K., and Jitendra B. Naik. 2016. "Optimization of Spray-Dried Diclofenac Sodium-Loaded Microspheres by Screening Design." *Drying Technology* 34 (13): 1593–1603. 10.1080/07373937.2016.1138121

Elham, Ghasemian, Parvizi Mahsa, Alireza Vatanara, and Ramezani Vahid. 2015. "Spray Drying of Nanoparticles to Form Fast Dissolving Glipizide." *Asian Journal of Pharmaceutics* 9 (3): 213–218. 10.4103/0973-8398.160319

Ferreira, Mónica P. A., João Pedro Martins, Jouni Hirvonen, and Hélder A. Santos. 2020. "Spray-Drying for the Formulation of Oral Drug Delivery Systems." *Nanotechnology for Oral Drug Delivery*, 253–284. 10.1016/b978-0-12-818038-9.00007-7

Fonte, Pedro, Francisca Araújo, Vítor Seabra, Salette Reis, Marco Van De Weert, and Bruno Sarmento. 2015. "Co-Encapsulation of Lyoprotectants Improves the Stability of Protein-Loaded PLGA Nanoparticles upon Lyophilization." *International Journal of Pharmaceutics* 496 (2): 850–862. 10.1016/j.ijpharm.2015.10.032

Fonte, Pedro, Paulo Roque Lino, Vítor Seabra, António J. Almeida, Salette Reis, and Bruno Sarmento. 2016. "Annealing as a Tool for the Optimization of Lyophilization and Ensuring of the Stability of Protein-Loaded PLGA Nanoparticles." *International Journal of Pharmaceutics* 503 (1–2): 163–173. 10.1016/j.ijpharm.2016.03.011

Fonte, Pedro, Salette Reis, and Bruno Sarmento. 2016. "Facts and Evidences on the Lyophilization of Polymeric Nanoparticles for Drug Delivery." *Journal of Controlled Release* 225: 75–86. 10.1016/j.jconrel.2016.01.034

Gülsün, Tuğba, R. Neslihan Gürsoy, and Levent Öner. 2009. "Nanocrystal Technology for Oral Delivery of Poorly Water-Soluble Drugs." *Fabad Journal of Pharmaceutical Sciences* 34 (1): 55–65.

Haggag, Yusuf A., and Ahmed M. Faheem. 2015. "Evaluation of Nano Spray Drying as a Method for Drying and Formulation of Therapeutic Peptides and Proteins." *Frontiers in Pharmacology* 6 (July): 1–5. 10.3389/fphar.2015.00140

Harsha, Sree, Bandar E. Al-Dhubiab, Anroop B. Nair, Mahesh Attimarad, Katharigatta N. Venugopala, and Kedarnath Sa. 2017. "Pharmacokinetics and Tissue Distribution of Microspheres Prepared by Spray Drying Technique: Targeted Drug Delivery." *Biomedical Research (India)* 28 (8): 3387–3396.

Hu, Jun, Yuancai Dong, Wai Kiong Ng, and Giorgia Pastorin. 2018. "Preparation of Drug Nanocrystals Embedded in Mannitol Microcrystals via Liquid Antisolvent Precipitation Followed by Immediate (on-Line) Spray Drying." *Advanced Powder Technology* 29 (4): 957–963. 10.1016/j.apt.2018.01.013

Ishwarya, S. Padma, C. Anandharamakrishnan, and Andrew G. F. Stapley. 2015. "Spray-Freeze-Drying: A Novel Process for the Drying of Foods and Bioproducts." *Trends in Food Science and Technology* 41 (2): 161–181. 10.1016/j.tifs.2014.10.008

Jacobs, Irwin C. 2014. "Atomization and Spray-Drying Processes." *Microencapsulation in the Food Industry*, 47–56. 10.1016/b978-0-12-404568-2.00005-4

Khairnar, Gokul, Vinod Mokale, Rajeshwari Khairnar, Arun Mujumdar, and Jitendra Naik. 2020. "Production of Antihyerglycemic and Antihypertensive Drug Loaded Sustained Release Nanoparticles Using Spray Drying Technique: Optimization by Placket Burman Design." *Drying Technology*: 1–12. 10.1080/07373937.2020.1825292

Khor, Chia Miang, Wai Kiong Ng, Kok Ping Chan, and Yuancai Dong. 2017. "Preparation and Characterization of Quercetin/Dietary Fiber Nanoformulations." *Carbohydrate Polymers* 161: 109–117. 10.1016/j.carbpol.2016.12.059

Li, Xiang, Nicolas Anton, Cordin Arpagaus, Fabrice Belleteix, and Thierry F. Vandamme. 2010. "Nanoparticles by Spray Drying Using Innovative New Technology: The Büchi Nano Spray Dryer B-90." *Journal of Controlled Release* 147 (2): 304–310. 10.1016/j.jconrel.2010.07.113

Li, Xiang, Nicolas Anton, Thi Minh Chau Ta, Minjie Zhao, Nadia Messaddeq, and Thierry F. Vandamme. 2011. "Microencapsulation of Nanoemulsions: Novel Trojan Particles for Bioactive Lipid Molecule Delivery." *International Journal of Nanomedicine* 6: 1313–1325. 10.2147/ijn.s20353

Liu, W., X. D. Chen, and C. Selomulya. 2015. "On the Spray Drying of Uniform Functional Microparticles." *Particuology* 22: 1–12. 10.1016/j.partic.2015.04.001

Maged, Amr, Azza A. Mahmoud, and Mahmoud M. Ghorab. 2017. "Hydroxypropyl-Beta-Cyclodextrin as Cryoprotectant in Nanoparticles Prepared By Nano-Spray Drying Technique." *Journal of Pharmaceutical Sciences & Emerging Drugs* 5 (1): 1–6. 10.4172/2380-9477.1000121

Meeus, Joke, David J. Scurr, Katie Amssoms, Martyn C. Davies, Clive J. Roberts, and Guy Van Den Mooter. 2013. "Surface Characteristics of Spray-Dried Microspheres Consisting of PLGA and PVP: Relating the Influence of Heat and Humidity to the Thermal Characteristics of These Polymers." *Molecular Pharmaceutics* 10 (8): 3213–3224. 10.1021/mp400263d

Mehta, Piyush, C. Bothiraja, Shivajirao Kadam, and Atmaram Pawar. 2018. "Potential of Dry Powder Inhalers for Tuberculosis Therapy: Facts, Fidelity and Future." *Artificial Cells, Nanomedicine, and Biotechnology* 46 (sup3): S791–S806. 10.1080/21691401.2018.1513938

Miller, Dave A., Daniel Ellenberger, and Marco Gil. 2016. *Spray-Drying Technology*. 10.1007/978-3-319-42609-9

Moayyedi, Mahsa, Mohammad Hadi Eskandari, Amir Hossein Elhami Rad, Esmaeil Ziaee, Mohammad Hossein Haddad Khodaparast, and Mohammad Taghi Golmakani. 2018. "Effect of Drying Methods (Electrospraying, Freeze Drying and Spray Drying) on Survival and Viability of Microencapsulated Lactobacillus Rhamnosus ATCC 7469." *Journal of Functional Foods* 40 (November 2017): 391–399. 10.1016/j.jff.2017.11.016

Momin, Mohammad A. M., Ian G. Tucker, and Shyamal C. Das. 2018. "High Dose Dry Powder Inhalers to Overcome the Challenges of Tuberculosis Treatment." *International Journal of Pharmaceutics* 550 (1–2): 398–417. 10.1016/j.ijpharm.2018.08.061

Mujumdar, Arun S. 2020. "Chemical Drying." *Handbook of Industrial Drying*.

Mukhopadhyay, Sayantan. 2018. *Nano Drugs: A Critical Review of Their Patents and Market. A Critical Review of Their Patents and Market. Characterization and Biology of Nanomaterials for Drug Delivery: Nanoscience and Nanotechnology in Drug Delivery*. Elsevier. 10.1016/B978-0-12-814031-4.00018-0

Naar, Z., N. Adányi, I. Bata-Vidács, R. Tömösközi-Farkas, and R. Tömösközi-Farkas. 2017. "Nanoencapsulation Technologies for the Food and Nutraceutical Industries S.M. Jafari Nanoencapsulation Technologies for the Food and Nutraceutical Industries 125 London Wall, London EC2Y 5AS, United Kingdom: Academic Press, Elsevier Science Publishing Co." *Acta Alimentaria* 46 (3): 390–394. 10.1556/066.2017.46.3.16

Nunes, Rafaela, Beatriz D. Avó Pereira, Miguel A. Cerqueira, Pedro Silva, Lorenzo M. Pastrana, António A. Vicente, Joana T. Martins, and Ana I. Bourbon. 2020. "Lactoferrin-Based Nanoemulsions to Improve the Physical and Chemical Stability of Omega-3 Fatty Acids." *Food and Function* 11 (3): 1966–1981. 10.1039/c9fo02307k

Pardeshi, Sagar, Pritam Patil, Rahul Rajput, Arun Mujumdar, and Jitendra Naik. 2020. "Preparation and Characterization of Sustained Release Pirfenidone Loaded Microparticles for Pulmonary Drug Delivery: Spray Drying Approach." *Drying Technology* 39 (3): 337–347. 10.1080/07373937.2020.1833213

Patil, Pritam, Gokul Khairnar, and Jitendra Naik. 2015. "Preparation and Statistical Optimization of Losartan Potassium Loaded Nanoparticles Using Box Behnken Factorial Design: Microreactor Precipitation." *Chemical Engineering Research and Design* 104: 98–109. 10.1016/j.cherd.2015.07.021

Pedrozo, Regiellen Cristina, Emilli Antônio, Najeh Maissar Khalil, and Rubiana Mara Mainardes. 2020. "Bovine Serum Albumin–Based Nanoparticles Containing the Flavonoid Rutin Produced by Nano Spray Drying." *Brazilian Journal of Pharmaceutical Sciences* 56: 1–8. 10.1590/s2175-97902019000317692

Pérez-Masiá, Rocío, Rubén López-Nicolás, Maria Jesús Periago, Gaspar Ros, Jose M. Lagaron, and Amparo López-Rubio. 2015. "Encapsulation of Folic Acid in Food Hydrocolloids through Nanospray Drying and Electrospraying for Nutraceutical Applications." *Food Chemistry* 168: 124–133. 10.1016/j.foodchem.2014.07.051

Piñón-Balderrama, Claudia I., César Leyva-Porras, Yolanda Terán-Figueroa, Vicente Espinosa-Solís, Claudia Álvarez-Salas, and María Z. Saavedra-Leos. 2020. "Encapsulation of Active Ingredients in Food Industry by Spray-Drying and Nano Spray-Drying Technologies." *Processes* 8 (8): 889. 10.3390/PR8080889

Salama, Alaa Hamed. 2020. "Spray Drying as an Advantageous Strategy for Enhancing Pharmaceuticals Bioavailability." *Drug Delivery and Translational Research* 10 (1): 1–12. 10.1007/s13346-019-00648-9

Sarabandi, Khashayar, Pouria Gharehbeglou, and Seid Mahdi Jafari. 2020. "Spray-Drying Encapsulation of Protein Hydrolysates and Bioactive Peptides: Opportunities and Challenges." *Drying Technology* 38 (5–6): 577–595. 10.1080/07373937.2019.1689399

Schuck, P. 2009. "Understanding the Factors Affecting Spray-Dried Dairy Powder Properties and Behavior." *Dairy-Derived Ingredients: Food and Nutraceutical Uses*: 24–50. 10.1533/9781845697198.1.24

Shishir, Mohammad Rezaul Islam, and Wei Chen. 2017. "Trends of Spray Drying: A Critical Review on Drying of Fruit and Vegetable Juices." *Trends in Food Science and Technology* 65: 49–67. 10.1016/j.tifs.2017.05.006

Shrimal, Preena, Girirajsinh Jadeja, Jitendra Naik, and Sanjaykumar Patel. 2019. "Continuous Microchannel Precipitation to Enhance the Solubility of Telmisartan with Poloxamer 407 Using Box-Behnken Design Approach." *Journal of Drug Delivery Science and Technology* 53 (April): 101225. 10.1016/j.jddst.2019.101225

Singh, Abhishek, and Guy Van den Mooter. 2016. "Spray Drying Formulation of Amorphous Solid Dispersions." *Advanced Drug Delivery Reviews* 100: 27–50. 10.1016/j.addr.2015.12.010

Sosnik, Alejandro, and Katia P. Seremeta. 2015. "Advantages and Challenges of the Spray-Drying Technology for the Production of Pure Drug Particles and Drug-Loaded Polymeric Carriers." *Advances in Colloid and Interface Science* 223: 40–54. 10.1016/j.cis.2015.05.003

Souza, Hugo Junior Barboza de, Anelise Lima de Abreu Dessimoni, Marina Letícia Alves Ferreira, Diego Alvarenga Botrel, Soraia Vilela Borges, Lívia Cássia Viana, Cassiano Rodrigues de Oliveira, Amanda Maria Teixeira Lago, and Regiane Victória de Barros Fernandes. 2020. "Microparticles Obtained by Spray-Drying Technique Containing Ginger Essential Oil with the Addition of Cellulose Nanofibrils Extracted from the Ginger Vegetable Fiber." *Drying Technology*: 1–15. 10.1080/07373937.2020.1851707

Sun, Wei, Rui Ni, Xin Zhang, Luk Chiu Li, and Shirui Mao. 2015. "Spray Drying of a Poorly Water-Soluble Drug Nanosuspension for Tablet Preparation: Formulation and Process Optimization with Bioavailability Evaluation." *Drug Development and Industrial Pharmacy* 41 (6): 927–933. 10.3109/03639045.2014.914528

Tania Marante, Claudia Viegas, Ines Duarte, Ana S. Macedo and Pedro Fonte. 2020. "An Overview on Spray-Drying of Protein-Loaded Polymeric Nanoparticles for Dry Powder Inhalation Tânia." *Pharmaceutics* 11 (11): 1–23. 10.3390/pharmaceutics12111032

Vehring, Reinhard. 2008. "Pharmaceutical Particle Engineering via Spray Drying." *Pharmaceutical Research* 25 (5): 999–1022. 10.1007/s11095-007-9475-1

Verma, Umakant, Arun Mujumdar, and Jitendra Naik. 2020. "Preparation of Efavirenz Resinate by Spray Drying Using Response Surface Methodology and Its Physicochemical Characterization for Taste Masking." *Drying Technology* 38 (5–6): 793–805. 10.1080/07373937.2019.1590845

Wagh, Pankaj, Arun Mujumdar, and Jitendra B. Naik. 2019. "Preparation and Characterization of Ketorolac Tromethamine-Loaded Ethyl Cellulose Micro-/Nanospheres Using Different Techniques." *Particulate Science and Technology* 37 (3): 347–357. 10.1080/02726351.2017.1383330

Wagh, Pankaj S., and Jitendra B. Naik. 2016. "Development of Mefenamic Acid–Loaded Polymeric Microparticles Using Solvent Evaporation and Spray-Drying Technique." *Drying Technology* 34 (5): 608–617. 10.1080/07373937.2015.1064947

Waghulde, Mrunal, and Jitendra Naik. 2017. "Comparative Study of Encapsulated Vildagliptin Microparticles Produced by Spray Drying and Solvent Evaporation Technique." *Drying Technology* 35 (13): 1645–1655. 10.1080/07373937.2016.1273230

Waghulde, Mrunal, Rahul Rajput, Arun Mujumdar, and Jitendra Naik. 2019. "Production and Evaluation of Vildagliptin-Loaded Poly(Dl-Lactide) and Poly(Dl-Lactide-Glycolide) Micro-/Nanoparticles: Response Surface Methodology Approach." *Drying Technology* 37 (10): 1265–1276. 10.1080/07373937.2018.1495231

Wais, Ulrike, Alexander W. Jackson, Tao He, and Haifei Zhang. 2016. "Nanoformulation and Encapsulation Approaches for Poorly Water-Soluble Drug Nanoparticles." *Nanoscale* 8 (4): 1746–1769. 10.1039/c5nr07161e

Wang, Taoran, Qiaobin Hu, Mingyong Zhou, Jingyi Xue, and Yangchao Luo. 2016. "Preparation of Ultra-Fine Powders from Polysaccharide-Coated Solid Lipid Nanoparticles and Nanostructured Lipid Carriers by Innovative Nano Spray Drying Technology." *International Journal of Pharmaceutics* 511 (1): 219–222. 10.1016/j.ijpharm.2016.07.005

Xue, Jingyi, Taoran Wang, Qiaobin Hu, Mingyon Zhou, and Yangchao Luo. 2018. "Insight into Natural Biopolymer-Emulsified Solid Lipid Nanoparticles for Encapsulation of Curcumin: Effect of Loading Methods." *Food Hydrocolloids* 79: 110–116. 10.1016/j.foodhyd.2017.12.018

Zahariev, Nikolay, Bissera Pilicheva, and Stanislava Simeonova. 2019. "Nano Spray-Drying: A Reliable Modern Approach for Drug Carriers Development" *Journal of Physics and Technology*3 (2): 48–55.

Zhang, Jiuhong, Zhiqiang Xie, Nan Zhang, and Jian Zhong. 2017. *Nanosuspension Drug Delivery System: Preparation, Characterization, Postproduction Processing, Dosage Form, and Application. Nanostructures for Drug Delivery*. Elsevier. 10.1016/b978-0-323-46143-6.00013-0

9 Superheated Steam Drying of Particulates

Sanjay Kumar Patel
Department of Mechanical Engineering, JSPM's Rajarshi Shahu College of Engineering, Pune, Maharashtra, India

Mukund Haribhau Bade
Department of Mechanical Engineering, Sardar Vallabhbhai National Institute of Technology, Surat, Gujarat, India

CONTENTS

9.1 INTRODUCTION

Superheated steam drying (SSD) is an emerging drying technology that involves a drying medium as superheated steam in a convective (direct) dryer in place of hot air or combustion/flue gases to supply thermal energy for drying/dewatering. The choice of drying medium depends on numerous parameters, including temperature and pressure of steam/gases, energy consumption, quality of the dried product, and performance of the drying medium (Mujumdar, 2014a; Patel and Bade, 2020). The energy consumption is higher for hot air drying intended for most drying applications,

DOI: 10.1201/9781003207108-9

especially in food, biomass, coal, chemical process industries, papermaking, industrial pulp, etc. (Mujumdar, 2014b). However, in the case of superheated steam as a drying medium, energy consumption can be reduced significantly, as discussed by Li et al. (1999) and Van Deventer and Heijmans (2001). SSD technique is becoming popular due to its potential applications and reuse of the exhausted surplus steam for drying. Hence, it has been proven to be an economical and environment-friendly technique in many industrial applications such as food and agricultural products, chemicals, papers, paint sludge, lumber, coal, and so forth (Mujumdar, 2014c). This chapter discussed the applicability of superheated steam for particulate matters. The above products are examples of particulate materials, considered hygroscopic porous materials. The pores of these particles are partially contained liquid water and partially filled with an air/water-vapor mixture. Further, the drying of these materials using a superheated steam in different dryers, drying fundamentals, and drying models are studied in detail. Moreover, the industrial tradition of steam drying, energy, and environmental aspects are discussed.

Experimental work related to superheated steam drying with measuring instrumentations, various experimental or lab scale work, and their potential have been discussed in this chapter. Then, theoretical work performed by multiple researchers is reviewed in detail with technological development. Furthermore, the mathematical model has been addressed to analyze the superheated steam drying process with moisture interaction of spherical particles and drying zone. In the next section, a comparison of hot air and superheated steam drying is given with the scope of operation. At the end of this chapter, key industries working in superheated steam drying with particulate materials having commercial products are highlighted.

9.1.1 Advantages and Limitations of SSD

Superheated steam is a viable option for drying because of several advantages, which improve the energy and carbon footprint of the process. Further, it is presently drawing attention in the food industries, medical treatments, and other industries where cleaning and disinfection play a crucial role. The advantages and limitations of superheated steam drying (SSD) are described by Mujumdar (2014c) as follows:

- Steam used is generally near ambient pressure; superheated and high-pressurized steam may not be adequate.
- Compared to hot air/gases, superheated steam has a high thermal capacity per unit volume and relatively high thermal conductivity.
- No oxidation or combustion reactions are possible in SSD because it is present in a gaseous state of water with low oxygen conditions. It prevents explosion hazards, lowers the possibility of fires, and products without any burns.
- SSD has higher energy efficiency if partial recirculation of exhaust steam can be done, and the remaining exhaust steam may be used either in processes or indirect heat recovery.
- Superheated steam has a much more powerful drying capability due to high thermal conductivity and heat transfer coefficients (Sehrawat, Nema, and Kaur, 2016).

- The higher thermal conductivity and heat capacity of superheated steam leads to higher drying rates for surface moisture above the so-called inversion temperature. Below the inversion temperature, drying in the hot air is faster. In the falling rate, the higher product temperature in SSD (over 100°C at 1 bar) and reduced diffusion resistance to water vapor (no air) lead to faster drying rates.
- Better pollution control by preventing gas emissions into the atmosphere is possible due to reuse and heat recovery, decent product quality, and food safety due to an oxygen-free operation (J. Li, Liang, and Bennamoun, 2016) compared with the conventional hot air drying techniques.
- SSD permits pasteurization, sterilization, and deodorization of food products.

Overall, SSD has reduced energy, carbon, and water footprints, making the process sustainable.

SSD has a few limitations that can be discussed in the following:

- A superheated steam drying system requires a more complex setup than a hot air/gas drying system. The generation of superheated steam requires a steam generator with a superheater system, which increases capital costs.
- SSD needs an excellent sealing system. Feeding and discharge of SSDs must not allow infiltration of air. It may increase the operating cost.
- Starting and closing are more complex operations for SSD than hot air/gases dryer.
- At the initial stage of SSD, condensation occurs before evaporation starts because the feed enters at the ambient temperature, and due to this, 10%–15% residence time in the dryer may increase.
- For drying heat-sensitive materials, specially designed low-pressure superheated steam drying is more appropriate and may also increase the drying rate.
- Products that may require oxidation reactions (enzymatic products like fruits, vegetables, seafood, etc.) to develop desired quality parameters (color, taste) cannot be dried by superheated steam except for non-enzymatic products. However, steam drying can be followed by an auxiliary heater with the help of air drying.
- The cost of the ancillaries (e.g., feeding systems, product collection systems, exhaust steam recovery systems, etc.) is typically much higher than the cost of the steam dryer alone. Hence, for the continuous operation of the SSD system, the techno economics of the ancillary equipment needs to be evaluated.
- It is important to note that there are currently limited experimental, pilot-scale studies and field experiences with SSD for a wide range of products.

9.2 OVERVIEW OF EXPERIMENTAL AND THEORETICAL STUDIES OF PARTICULATE DRYING

The importance of experimentation and theoretical work for drying particulates is described in detail by Kemp and Oakley (2002). In this seminal work, the authors proposed the challenges in drying particulate matters than other liquids and vapor phase processes. Furthermore, experimental validation is proposed for theoretical analysis, at least with a pilot study, to get the practically feasible results as specific data need to be deduced with the help of experiments only. To determine drying kinetics, the specific properties of particulate matter are essential; however, they are highly dependent on the characteristics and structure of the solid and the upstream particle formation process; hence, they must be evaluated experimentally. The review of different methods for measuring drying kinetics of loose and particulate material is given in the textbook of Keey (1992). Several researchers proposed various experimental techniques to investigate the drying kinetics of single particles; a few of them are conventional microbalance (Hirschmann and Tsotsas, 1998), acoustic levitator (Groenewold et al., 2000), drying tunnel (Groenewold, Groenewold, and Tsotsas, 2000), and magnetic suspension balance proposed by Kwapinski and Tsotsas (2006). Looi et al. (2002) performed several experiments to investigate the drying kinetics of single porous particles of ceramic and lignite using pressurized superheated steam. For the SSD, the equilibrium moisture content is higher for high-pressure steam; hence, the equilibrium condition is reached earlier, but a rise in pressure does not affect the drying rate for lignite particles. Experimental methods proposed above for kinetic analysis depend on elements like material and vapor properties, operating conditions (flow velocity, temperature, pressure), and required accuracy, as Kwapinski and Tsotsas (2006) discussed. The authors proposed magnetic suspension balance, a novel technique for assessing single-particle drying kinetics of particulate materials, which is stable, well reproducible for experiments with high gas flow velocities around the particle, and wide operational ranges can be observed for temperature and particle diameter. Overall, it is noted that the previously discussed experimental techniques (conventional microbalance, acoustic levitator, and drying tunnel) are restricted to the atmospheric conditions, and have constraints based on operating parameters as flow, temperature, and pressure of drying process. These restrictions can be removed by single-particle measurements in a magnetic suspension balance to a great extent. Makkawi and Ocone (2009) experimentally determined the overall mass transfer coefficient in a conventional bubbling fluidized bed dryer by using an electrical capacitance tomography (ECT) imaging sensor for hot air as a working fluid. The ECT sensor is helpful to provide dynamic statistics on the fluidized bed material distribution and it allows quantifying the bubble diameter and velocity only. However, exhaustive experimentation needs to be conducted for wider operating conditions (particle size, gas velocity, water content, porous/nonporous particles). Further, it is important to check the compatibility of a superheated steam dryer with ECT for experimental purposes.

In recent years, proven theoretical studies and simulation of drying techniques have been considered essential tools for design and operational control; however, most of the work is for hot air as a working fluid instead of superheated steam. In theoretical analysis, due to superheated steam as a working fluid, heat transfer from the superheated steam to the drying feed/material takes place by convection, and evaporation of surface/bound moisture is due to boiling of water only instead of mass transfer. A two-fluid continuum model for the drying of particulate materials (PVC and sand particles) using a pneumatic dryer is presented by Levy and Borde (2001).

After validation of the model, it is compared with the models proposed in the literature. The proposed two-fluid continuum pneumatic dryer model for drying particulate materials shows good agreement with DryPak (Gong and Mujumdar, 2008).

Defo et al. (2004) developed a mathematical model based on a two-dimensional finite element method for vacuum-contact drying using superheated steam as a drying media. Further, simulation results are compared with experimental work obtained from superheated steam vacuum drying for sugar maple sapwood, found very close to each other when transfer coefficients are adjusted as a function of wood moisture content. The book chapter by Mujumdar (Mujumdar, 2006) explained the various dryers currently in use with a detailed area of application for better selection and advanced drying techniques. Researchers are presently working on particulate materials. For drying pumpable liquids (solutions, suspensions, or slurries), thin or thick pastes (including sludge), etc., spray and drum dryers are generally preferred to form particulate material nowadays. For drying granular or particulate solids, the most common dryers in use today are cascading rotary dryers with or without internal steam tubes, conveyor dryers, and continuous tray dryers (e.g., turbo or plate dryers), which must compete with fluidized-bed dryers (with or without internal exchangers) and vibrated bed dryers, among others. Impinging streams, rotating spouted beds, pulsed-fluid beds, etc., are the latest drying technologies explored by the researchers (Mujumdar, 2006).

Ortega-Rivas (2012) discussed that particulate food materials tend to be soft and friable, while their physicochemical nature may cause a release of sticky substances or a tendency to absorb moisture. These characteristics complicate physical determinations and behavior under handling and processing conditions. Furthermore, the author proposed methods to measure various properties of food particulates. For impinging stream dryer (ISD) with air as working fluid, realizable k-ε turbulence models performed better than standard k-ε turbulence models (Choicharoen, Devahastin, and Soponronnarit, 2012). A modified numerical model can predict the behavior of lignite particles ranging from a few millimeters to several centimeters in size and dried at temperatures from 383 K to 443 K (110°C–170°C). Hence, it is suitable for simulations of drying behaviors of lignite with superheated steam when an appropriate heat transfer coefficient is given (Kiriyama et al., 2014).

More recently, Le et al. (2020) presented a multiscale model for envisaging the superheated steam drying behavior and characteristics of a ceramic material in a packed bed dryer by using a reaction engineering approach (REA). The authors compared the results of the REA model to both experimental and numerical simulations of Chen et al. (2000) and Woo et al. (2013) and found a more accurate drying behavior of SSD. At higher drying temperatures of 150°C and 175°C, the drying rate curves obtained from both models, the REA and Chen, are close to the experimental observation. Overall, the study confirmed that the REA model could be considered a simple and effective model to describe the SSD process because of its predictability and extrapolative ability in a wide range of bulk temperatures and velocities. The above articles show the potential of research activities in particulate drying using superheated steam drying media. But to date, the technology is not mature and is fully commercialized.

9.2.1 Discussion on Various Drying Techniques for Particulates Using SSD

For drying particulate solids, many well-proven techniques, such as fluidized bed dryer, spouted bed dryer, pneumatic and flash dryer, rotary dryer, conveyor dryer, etc., are employed for hot air as working fluid, have also been tried with superheated steam. While drying with superheated steam, certain modifications are needed in the properties of the drying media (superheated steam). As the viscosity of the superheated steam is lower than the hot air, the residence time will be shortened, and the height/length of the drying chamber needs to be adjusted accordingly. Furthermore, many difficulties come up during the drying of paste-like materials and suspensions of high moisture content due to their consistency. For example, overheating and crusting affect the quality of the product. While drying with superheated steam, certain modifications in the dryer are needed compared to conventional hot-air dryers due to changes in the properties.

9.3 SUPERHEATED STEAM FLUIDIZED BED DRYER

Fluidized bed dryers (FBDs) are used extensively to dry wet particulate and granular materials in the particle size 50–5,000 μm (Law and Mujumdar, 2014). To evolve a greater understanding of steam fluidized bed drying, a few researchers examined steam drying of 2–30 mm single lignite particles (Kiriyama et al., 2013, 2014). Further, they developed a mathematical model for a single lignite particle and compared it with experimental results to determine their drying characteristics in constant and falling rate drying periods. Due to a higher temperature of steam, cracks on particles are observed. The heat transfer on the surface of the lignite particle depends on particle size and temperature. In another study, Kiriyama et al. (2016) studied the drying rate of single lignite particles in superheated steam with particle sizes having 100, 50, and 6 mm in diameter. Sesso and Franks (2017) developed a technique for testing saturated and dry particulate materials comprised of colloidal-sized particles and investigated the effect of saturation on fracture toughness. The influence of a hygroscopic porous particle (silica gel beads) used as the fluidizing particle in superheated steam fluidized bed dryer under atmospheric pressure is demonstrated by Tatemoto et al. (2008), and the results were compared with an inert particle (spherical glass beads) fluidized bed. Relatively, in the case of superheated steam, it is found that the drying completes earlier in the case of silica gel beads than in the case of glass beads. The water transfer from the sample to the fluidizing particles influenced the drying characteristics and observed drying time longer in the case of superheated steam than the hot air under the given operating conditions. The surface evaporation rate is higher in the case of hot air than in superheated steam when the drying gases/air temperature is lower than the inversion temperature (below 180°C–200°C). Overall, it is observed that the drying time is shorter in the case of hygroscopic porous particles than in the case of an inert particle in all the cases in the study.

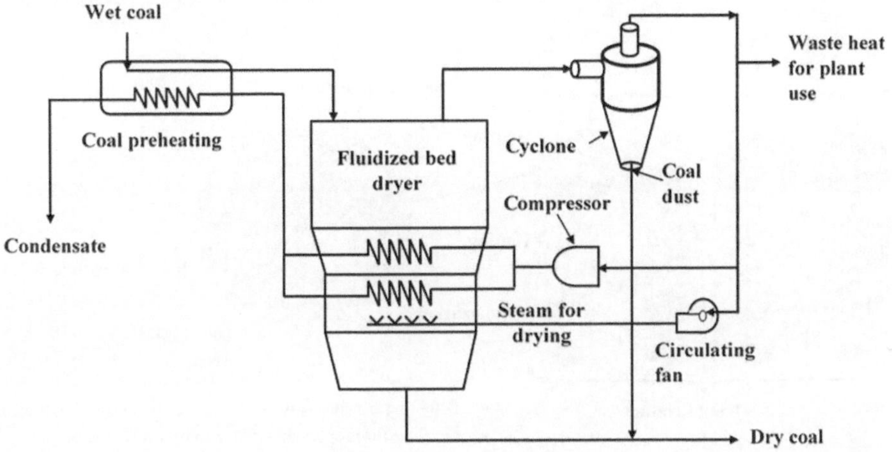

FIGURE 9.1 Schematic diagram of WTA superheated steam fluidized bed dryer (Jangam, Karthikeyan, and Mujumdar, 2011).

The most advanced technique for drying coal particles using a steam fluidized bed dryer is WTA (Wirbelshicht-Trocknungmit-interner, a German abbreviation for fluidized-bed drying with internal waste-heat utilization), firstly established by Germany on a commercial scale as reported by Jangam et al. (2011). The main features of the WTA process (Figure 9.1) include a fluidized bed dryer using superheated steam, vapor compression for recovering the latent heat from the process, and supply of heat energy to the drying solids. The coal is dried from around 60% moisture content to 12% using steam at 110°C at a low pressure of 50 mbar. A part of the steam at a higher temperature is indirectly used to heat the fluidizing bed through submerged tube bundles. The authors observed that the WTA process consumes 80% less energy than a rotary steam tube dryer, with 80% less dust emissions and lower capital investment (Karthikeyan, Zhonghua, and Mujumdar, 2009). The influences of the various parameters like drying temperature, pressure, coal size, and recent developments in drying and dewatering technologies are critically reviewed for optimal performance (Rao et al., 2015; Zhu, Wang, and Lu, 2015). Compared to rotary drying, fluidized bed drying, and microwave drying, the studies show that mechanical/thermal dewatering (MTE) drying is more efficient because of lower energy consumption.

9.4 PARTICULATE DRYING MECHANISM USING SSD

The superheated steam dryer (SSD) provides a novel configuration for drying particulate materials or paste or suspensions to obtain dried solid products. Two processes occur concurrently in the SSD technique as a moist/wet product passes off steam drying. The first process is transferring energy from the superheated steam to evaporate the surface moisture. In this process, humid/damp material (feed) enters at ambient temperature and comes in contact with superheated steam; under

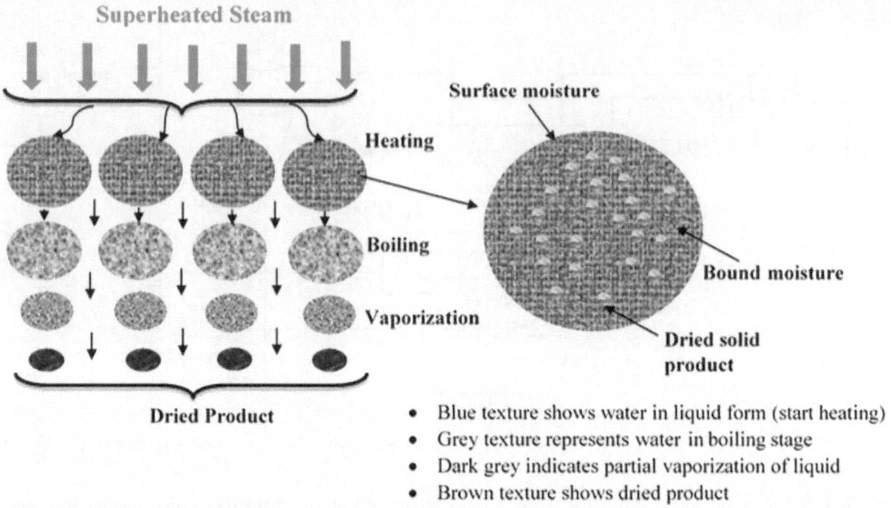

FIGURE 9.2 Drying mechanism of particulate materials.

this condition, there is unavoidable condensation that begins before drying, which increases drying time significantly. This problem is appropriately addressed either through local heating or preheating of feed. After that, the feed temperature rises to phase change temperature (100°C), beginning the constant drying rate period to remove the surface moisture. The second process involves the transfer of internal moisture from a solid to its surface; after that, similar to the first process, surface moisture will be evaporated (Mujumdar, 2014b), as represented in Figure 9.2. Drying of food and agricultural products (fruits and vegetables), chemicals, coal, and many more are examples of particulate drying.

A particulate material was assumed to initially consist of surface moisture, bound moisture, and solid product uniformly distributed over the volume of the particle, as represented in Figure 9.2. Surface moisture is known as water present on the outer surface of the product that evaporates due to sensible and latent heat under the given pressure in the case of the superheated steam. In this stage, because of the vaporization of surface water, a slight shrinkage of the particle might be observed (Figure 9.2).

After that, the bound moisture is uniformly distributed in solid particles over the volume that usually evaporates above the boiling point (greater than 100°C). In this constant drying period, capillary forces have to carry moisture from the inside of the solid to the surface, and the drying rate may still be stable. It is due to the continuous diffusion of the water inside the solid to its outer surface. The surface moisture transferred from the internal solid evaporates due to the latent heat of the superheated steam. In this case, the evaporation rate is lower than the diffusion rate of water transferred from the inside, so the water film is maintained at the solid surface. On further drying of the solid, the rate at which moisture may move through the solid due to concentration gradients between the inside and the outer surface of the material may be reduced continuously, and it may be categorized as a falling drying period, and generally starts from critical moisture content. There will be conduction heat transfer of the solid, which increases the temperature of the solid from the saturated temperature of the steam. The moisture diffusion controls the drying rate from the inside to the surface and then evaporation transfer from the surface. During this stage, some of the moisture bound by sorption is removed. As the moisture concentration is lowered by drying, the rate of internal movement of moisture decreases. The drying rate falls even more rapidly than before and continues until the moisture content falls to the equilibrium value for the prevailing air humidity, and then drying stops.

9.5 HEAT AND MASS TRANSFER PHENOMENA IN PARTICULATE DRYING

Superheated steam has heat transfer properties superior to hot air at the same temperature. Since there is no resistance to diffusion of the evaporated water in its own vapor, the drying rate in the constant rate period is dependent only on the heat transfer rate. The convective heat transfer coefficient (h) between the steam and the solid surface is depicted in Figure 9.3, and it can be estimated using standard correlations for interphase heat transfer. Neglecting heat transfer, heat losses, and other modes of heat transfer, the rate at which surface moisture evaporates into steam is given as Equation (9.1) (Kudra and Mujumdar, 2009):

$$\Delta Q = hA_s(T_{SS} - T_s) \tag{9.1}$$

where ΔQ is the rate of change of heat transfer, (W); h is the convective heat transfer coefficient, (W/m^2 °C); A_s is the heat transfer area, (m^2); T_{ss} is the temperature of superheated steam; and T_s is the surface temperature of particles, (°C), taken as the saturated temperature of steam (superheated steam at atmospheric pressure, surface temperature 100°C).

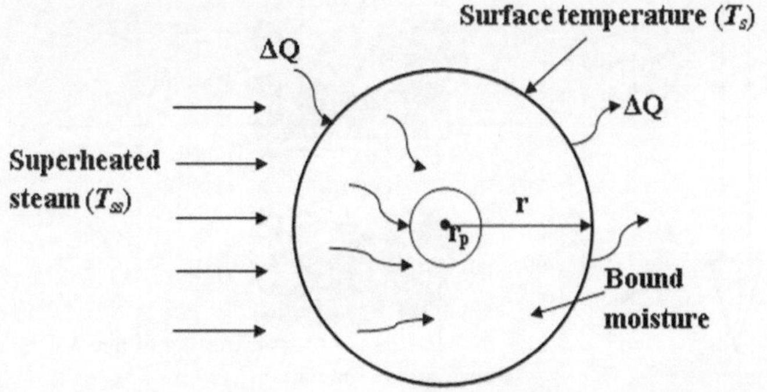

FIGURE 9.3 Heat and mass transfer in a single particle of particulate material.

9.5.1 Estimation of Moisture Content and Drying Rate

Moisture content is the quantity of water contained in the feed particle. It is expressed as a percentage of moisture based on the total weight (wet basis), X_w or dry matter (dry basis), X_d. The moisture content is expressed as

$$X_w = \frac{w - d}{w} \times 100 \tag{9.2}$$

$$X_d = \frac{w - d}{d} \times 100 \tag{9.3}$$

where X_w and X_d are the moisture content on a wet and dry basis, respectively; subscript w is the total weight (also called wet weight); and subscript d is the dry weight. The drying rate is the ratio of a difference between the initial and final weight to the time interval.

The behaviour of solid particles can be characterized by measuring the loss of moisture content as a function of time. Kiriyama et al. (2014) depict a typical drying rate of lignite coal (Figure 9.4) from initial to equilibrium moisture content. It is observed that during the constant rate period, the heat transfer dominates the drying process, and in this period, a large heating surface is preferred. However, during the falling or decreasing rate period, the internal resistances control the drying rate, and this period needs a longer residence time. The thermal

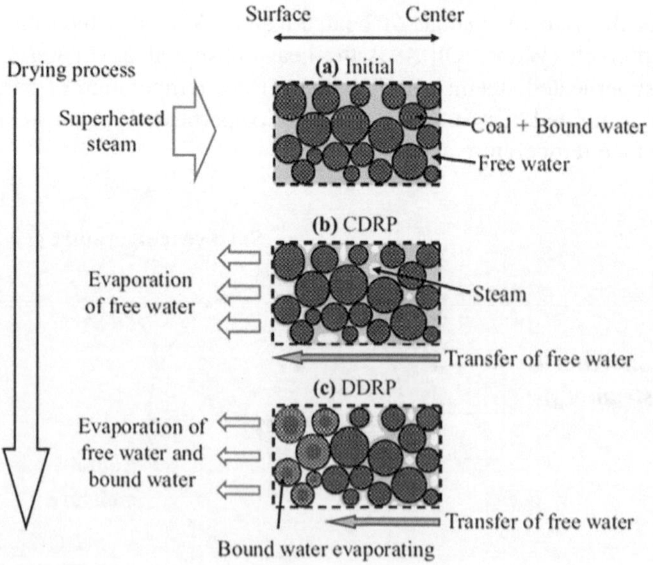

FIGURE 9.4 Illustration of drying rate (a) initial condition, (b) constant drying rate period, and (c) decreasing drying rate period for lignite coal taken from Kiriyama et al. (2014).

energy transfer into the particle for a constant rate period is evaluated by using Equation (9.4):

$$\Delta Q = X_{rr} \times L_{vap} \tag{9.4}$$

where L_{vap} is the latent heat of vaporization at saturation temperature (J/kg).

Simplifying Equations (9.1) and (9.4), time for constant rate period (t_c) is evaluated, and it is expressed by Equation (9.5) as follows, if considering heat transfer between the particle using superheated steam:

$$X_{rr} = \frac{x_0 - x_c}{t_c} = \frac{hA_s}{L_{vap}}(T_{ss} - T_s) \tag{9.5a}$$

$$t_c = \frac{L_{vap} \times (x_0 - x_c)}{hA_s(T_{ss} - T_s)} \quad t_c = \frac{L_{vap} \times (x_0 - x_c)}{4\pi r^2 h(T_{ss} - T_s)} \tag{9.5b}$$

where X_{rr} is the moisture removal rate (kg water/s-kg dry solid); x_O is the initial moisture content (kg water/kg dry solid); x_c is the critical moisture content (kg water/kg dry solid); and t_c is the time of constant rate period, (s).$A_s = 4\pi r^2$ Area of the sphere (solid particle) and r is the radius of a droplet.

A liquid or vapor diffusion mechanism controls the drying phenomenon in the falling rate period. Fick's second law gives the relationship between the diffusion coefficient and the concentration gradient in a given environment. The time for falling rate period (t_f) for a spherical particle (drying material) is evaluated by Equation (9.6), which is taken from Mujumdar (2014a) as follows:

$$t_f = \frac{r_p^2}{\pi^2 D_{eff}} \ln\left[\frac{6}{\pi^2}\left(\frac{x_c - x_e}{x - x_e}\right)\right] \tag{9.6}$$

where r_p is the radius of a particle at the critical point; D_{eff} is the effective diffusion coefficient (m^2/s) of material and is the function of temperature, but it can be considered constant because temperature changes were experimentally found to be insignificant (Keey, 2014); x_c is the moisture content at the critical point between constant and falling rate period; x_e is the moisture content at equilibrium; and x is the local moisture content. So, the total time required for drying droplets in the dryer is expressed as Equation (9.7):

$$t = \frac{L_{vap} \times (x_0 - x_c)}{4\pi r^2 h(T_{ss} - T_s)} + \frac{r_p^2}{\pi^2 D_{eff}} \ln\left[\frac{6}{\pi^2}\left(\frac{x_c - x_e}{x - x_e}\right)\right] \tag{9.7}$$

9.6 COMPARATIVE STUDIES OF PARTICULATE DRYING USING SSD AND HOT AIR

This section discusses comparative research between hot air and superheated steam drying for particulate materials. Choicharoen et al. (2011) performed a comparative

TABLE 9.1
Comparative Results of Hot Air and Superheated Steam Impingement Drying System at Various Operating Conditions (Adapted from Choicharoen et al. (2011))

Drying Conditions				Hot Air Drying		Superheated Steam Drying	
V (m/s)	T (°C)	L (cm)	W_p ($kg_{dry\ solid}$/h)	Volumetric Water Evaporation Rate (kg_{water}/m^3h)	Volumetric Heat Transfer Coefficient (W/m^3k)	Volumetric Evaporation Rate (kg_{water}/m^3h)	Volumetric Heat Transfer Coefficient (W/m^3k)
20	130	5	10	308	3012	166	4561
			20	423	4407	228	7016
		13	10	228	2550	123	3323
			20	362	4079	196	5729
	150	5	10	351	3090	276	4523
			20	456	4446	397	7049
		13	10	281	2576	211	3399
			20	401	4106	328	5718
	170	5	10	398	3123	394	4585
			20	493	4426	562	7034
		13	10	332	2613	293	3370
			20	436	4105	463	5722
	190	5	10	–	–	505	4559
			20	–	–	723	7015
		13	10	–	–	373	3341
			20	–	–	596	5712

analysis of hot air and superheated steam impinging stream dryer for high moisture particulates materials (soy residue). The performance of the drying system, in terms of volumetric evaporation rate and volumetric heat transfer coefficient, was assessed at different operating conditions, which are summarized in Table 9.1. It was observed that the volumetric water evaporation rate of the superheated steam impingement stream drying (SSISD) system is considerably higher at an above drying temperature of 170°C. However, at below drying, a temperature of 150°C, the volumetric water evaporation rate is more significant in the case of hot air impingement stream drying (HAISD) in all the circumstances. It may be because the inversion temperature lies between 150°C and 170°C. The highest volumetric water evaporation rate is found at 807 kg water/m^{3}h in higher velocity (27 m/s), drying temperature (190°C), impinging distance (13 cm), and material feed flow rate (20 kg dry solid/h) in the case of superheated steam drying. In addition, the volumetric heat transfer coefficient is more significant in superheated steam drying than hot air drying at the same corresponding conditions in all the cases. Stokie et al. (2013) compared the drying characteristics of three types of lignite particles in the steam-fluidized and air-fluidized beds. The results obtained from the experiments show that both air and steam-fluidized bed drying results are similar. However, the coal particle dried faster than the steam-fluidized bed in the air-fluidized bed. Celen and Erdem (2018) conducted experiments to investigate the effect of hot air and low-pressure steam on the drying rate of the Turkish lignite of

30 mm particle size. The experiments showed that the surface and center temperatures of lignite particle drying with superheated steam increase rapidly compared to hot air. Moreover, the drying rate of SSD is higher than hot air drying at the beginning of the drying process due to higher mass transfer. While, after the condensation of water on the lignite particle, the drying rate with hot air is greater.

In general, it was observed that both (hot air and superheated steam) drying media have their own merits and demerits. The performance of drying media mainly depends on the type of application. It can be evaluated by considering the heat and mass transfer in a product, moisture removal rate, geometrical variations (shrinkage, dilation, and deformation), and physical and micro-structural modifications (color, flavor, appearance, nutrients, germination, depending on the product species, etc.). Moreover, the particulate matters are water-absorbing material (hygroscopic), so more attention needs to be taken during water evaporation.

9.7 INDUSTRIES INVOLVED IN MANUFACTURING OF SSD FOR PARTICULATE MATERIALS

Drying with superheated steam is an emergent technology with many potential advantages, such as energy savings, emissions reduction, fire and explosion prevention, product quality, etc. However, the industrialization of this technique is limited and immature for significant products due to a lack of research. In the open literature (research papers, conference articles, magazines, reports), including websites, the authors found very limited manufacturers, industries, and research organizations that work in the SSD. The leading manufacturers of industrial superheated steam dryers worldwide are listed as follows:

1. Hauni Maschinenbau AG, Hamburg, Germany
2. BHS-Sonthofen (India) Pvt. Ltd., Erramanzil/Banjara Hills, Hyderabad 500 034, India/BHS-Sonthofen GmbH, An der Eisenschmelze 47, 87527 Sonthofen, Germany
3. Shandong Tianli Energy Co., Ltd., No. 2000 Shunhua Road, High-tech Zone, Jinan, Shandong Province, 250101, China
4. Swedish Exergy AB, Gamla Rambergsvägen 34, SE-417 10, Gothenburg, Sweden/Exergy Dryers Pvt Ltd, Surajpur, Greater Noida, Dist. Gautambudh Nagar, Pin – 201306 (U.P.), India
5. Mitsui E&S Machinery Co., Ltd., Chuo-ku, Tokyo 104–8439, Japan
6. Tsukishima Kikai Co., Ltd., Harumi, Chuo-ku, Tokyo104–8439, Japan
7. SSP Private Limited, Faridabad, Haryana 121003, India
8. Louisville Dryer Company, Louisville, Kentucky 40219, United States

In addition to that, many institutes and universities are working on SSD. However, analytical, numerical, and experimental work is still limited in the open domain under the heading of particulate materials drying. In the following subsection, some SSD equipment available in the market is discussed. These apparatuses are specifically used to dry sludge, coal, pulp, lumber, paper, tissues, wood particle, and wood wafers. These dryers function continuously, as a large amount of water needs to be evaporated.

9.7.1 SSD Techniques Available on an Industrial Scale

9.7.1.1 GEA Exergy Barr-Rosin Dryer

The GEA exergy dryer was initially developed at Chalmers University in Sweden. At the beginning of the 1990s, it was made over to NIRO, now part of the GEA/ Barr-Rosin Company, Gothenburg, Sweden (Van Deventer, 2004). This type of dryer is also called a flash dryer (Van't Land, 2012). The dryer is operated slightly above atmospheric pressure as the usability of the steam increases with its pressure in the range of 7–20 bar, temperature 211°C, the velocity of steam in the dryer, and the heat exchanger is 20–40 m/s; and the residence time of solid particles is in the range of 5–60 s. This drying system can be helpful for the drying of solid materials like paper pulp, sawdust, wood waste, animal feed, wet distillers grain, peat, sludge, tea, agro biomass, bagasse, sugar beet pulp, wet solid waste, etc.

The working principle is based on a flash dryer, where a fan circulates pressurized superheated steam with feed. The thermal energy is transferred from the tube walls to the circulating steam (an external heater is provided around the tubes), where sensible heat is converted into latent heat. This latent heat is used to evaporate water from the moist feed. The dried product is separated from the circulating steam in a cyclone separator and taken out through a pressure-tight rotary valve, as shown in Figure 9.5.

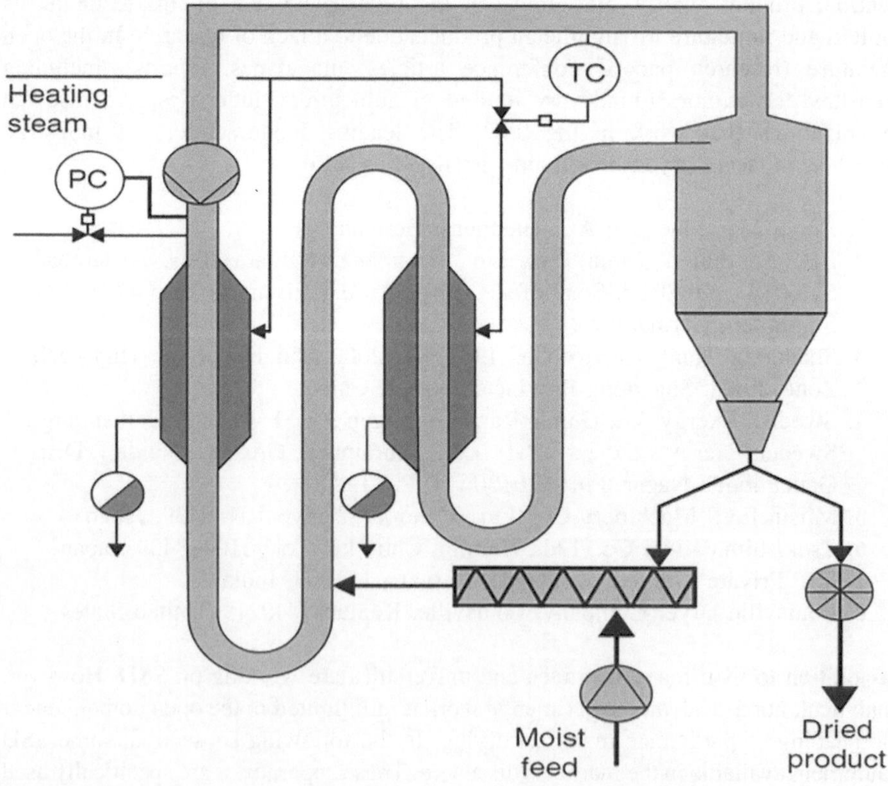

FIGURE 9.5 Superheated steam flash dryer (GEA Exergy Barr-Rosin Dryer) adapted from Land (Van't Land, 2012).

9.7.1.2 Pressurized Superheated Steam Fluidized Bed Dryer (BMA/NIRO)

The pressurized superheated steam fluidized bed dryer was first developed by EnerDry (1981) in Denmark for the sugar industry with an immersed heat exchanger. Later on, the equipment was commercialized for beet pulp drying, and further, the technology was adopted by other areas like the USA, Western Europe, Japan, etc. The dryer is working below 210°C temperature, 3.5 bar pressure, and its evaporating capacity is 5–70 t/h. The illustrative design of the dryer is shown in Figure 9.6. This drying system can also be used for other products like spent grain from ethanol plants, starch plants, breweries, sludge from wastewater, etc.

The pulp is fed through the inlet rotary valve (1) to the screw conveyor (2), bringing the pulp into the pressure vessel (3), which is pressurized with superheated steam of up to 3 bar. The only moving part in the dryer is the fan (4), circulating the steam through the heat exchanger (12), up through the perforated curved bottom (5), and into the low ring-shaped fluid bed (6).

As the arrows indicate, the pulp is kept "fluid" by swirling around in the fluid bed. Guiding plates force the pulp to move forward in the fluid bed until it arrives at

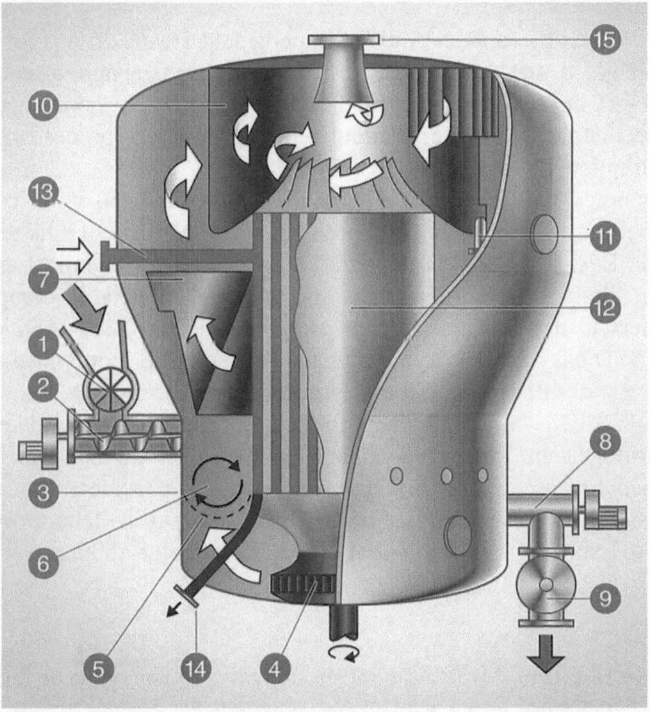

FIGURE 9.6 Illustrative diagram of pressurized fluidized bed superheated steam dryer; (1) rotary valve; (2) change screw; (3) pressure vessel; (4) impeller; (5) perforated curves bottom; (6) ring-shaped fluid bed; (7) plates; (8) discharge screw; (9) rotary valve; (10) cyclone; (11) dust outlet pipe; (12) heat exchanger; (13) steam inlet pipe; (14) condensed water outlet pipe; (15) exhaust outlet (Jensen, 1981).

the outlet conveyor (8) and leaves the dryer as dried pulp through the outlet rotary valve (9). Due to the reduced velocity of the superheated steam in the conical part of the pressure vessel, the lighter particles suspended in the steam fall down onto the forward inclined plates and slide forward in the pressure vessel. In this way also, the lighter particles pass along the dryer and are discharged through the outlet rotary valve. The circulating steam passes through the upper cylindrical part into the primary cyclone (10), where fine dust still suspended in the steam is separated from the steam. The dust utilizes an ejector led through a pipe (11) into the outlet conveyor. The dust is carried out of the pressure vessel together with the dried pulp.

From the cyclone, the dust-free steam passes down through the tubes in the heat exchanger (12), where it is reheated by the steam of a higher pressure supplied through the pipe (13) to the primary side of the heat exchanger. The supply steam is condensed in the heat exchanger and leaves the dryer through the condensate outlet (14). A higher supply steam pressure will heat the circulating steam to a higher temperature, which will increase the dryer's capacity. The steam evaporated from the pulp leaves through a pipe (15) and is recirculated back after reheating up to initial conditions.

9.8 CONCLUSIONS

This chapter studies the applicability of superheated steam drying for particulate materials. It is noted that the superheated steam drying is enabling technology for sustainability and improved product quality as there is no emissions mainly due to exhaust steam that may be either recirculated or used in the process, reducing the energy consumption.

Also, it is noted that a reaction engineering approach (REA) is a simple and effective model for describing the SSD process in a wide range of bulk temperatures and velocities. Besides inversion temperature, physical and chemical aspects such as shrinkage, dilation, browning, hardening, and loss of volatile components are vital parameters to decide the applicability of hot air and SSD. Above inversion temperature, SSD has a higher drying rate and proved more economical than hot air dryers; however, it will never be the sole criteria for selection.

Even though SSD is commercialized for drying significant particulates products, there is insufficient information available in the open domain and much less awareness than hot air dryers. So, it is necessary to provide more attention to quantitative and qualitative studies. Further, there needs to be laboratory scale and pilot studies to explore its full potential and wide range of applications.

REFERENCES

Celen, P., and H. H. Erdem. 2018. "An Experimental Investigation of Single Lignite Particle Dried in Superheated Steam and Hot Air." *International Journal of Coal Preparation and Utilization*: 1–10. doi:10.1080/19392699.2018.1536047.

Chen, Z., W. Wu, and P. K. Agarwal. 2000. "Steam-Drying of Coal. Part 1. Modeling the Behavior of a Single Particle." *Fuel* 79 (8): 961–974. doi:10.1016/S0016-2361(99)00217-3

Choicharoen, K., S. Devahastin, and S. Soponronnarit. 2011. "Comparative Evaluation of Performance and Energy Consumption of Hot Air and Superheated Steam Impinging Stream Dryers for High-Moisture Particulate Materials." *Applied Thermal Engineering*.

Choicharoen, K., S. Devahastin, and S. Soponronnarit. 2012. "Numerical Simulation of Multiphase Transport Phenomena During Impinging Stream Drying of a Particulate Material." *Drying Technology* 30 (11–12): 1227–1237. doi:10.1080/07373937.2012.704467

Defo, M., Y. Fortin, and A. Cloutier. 2004. "Modeling Superheated Steam Vacuum Drying of Wood." *Drying Technology* 22 (10): 2231–2253. doi:10.1081/DRT-200039984

Deventer, H. C. Van. 2004. "Industrial Superheated Steam Drying." *TNO Environment, Energy and Process Innovation.*

Deventer, H. C. Van, and R. M. H. Heijmans. 2001. "Drying with Superheated Steam." *Drying Technology* 19 (8): 2033–2045. doi:10.1081/DRT-100107287

Gong, Z. X., and A. S. Mujumdar. 2008. "Software for Design and Analysis of Drying Systems." 26 (7): 884–894. Taylor & Francis Group. doi:10.1080/07373930802142390 Http://Dx.Doi.Org/10.1080/07373930802142390

Groenewold, C., H. Groenewold, and E. Tsotsas. 2000. "Interrelations between Porous Structure and Convective Drying Kinetics in Theory and Experiment." In *12th International Drying Symposium*, 58.

Groenewold, C., C. Moser, H. Groenewold, and E. Tsotsas. 2000. "Determination of Single-Particle Drying Kinetics in an Acoustic Levitator." In *12th International Drying Symposium*, 28–31.

Hirschmann, C., and E. Tsotsas. 1998. "Impact of the Pore Structure on Particle Side Drying Kinetics." In *11th International Drying Symposium*, 216–223.

Jangam, S. V., M. Karthikeyan, and A. S. Mujumdar. 2011. "A Critical Assessment of Industrial Coal Drying Technologies: Role of Energy, Emissions, Risk and Sustainability." *Drying Technology* 29 (4): 395–407.

Jensen, A. S. 1981. "Steamdrying of Beet Pulp." *EnerDry*.

Karthikeyan, M., W. Zhonghua, and A. S. Mujumdar. 2009. "Low-Rank Coal Drying Technologies—Current Status and New Developments." *Drying Technology* 27 (3): 403–405.

Keey, R. B. 1992. *Drying of Loose and Particulate Materials*. New York: CRC Press: Taylor & Francis Group.

Keey, R. B. 2014. "Drying of Fibrous Materials." In *Handbook of Industrial Drying*, Fourth ed., 777–802. Boca Raton; London; New York: CRC Press: Taylor & Francis.

Kemp, I. C., and D. E. Oakley. 2002. "Modelling of Particulate Drying in Theory and Practice." *Drying Technology* 20 (9): 1699–1750. doi:10.1081/DRT-120015410

Kiriyama, T., H. Sasaki, A. Hashimoto, S. Kaneko, and M. Maeda. 2013. "Experimental Observations and Numerical Modeling of a Single Coarse Lignite Particle Dried in Superheated Steam." *Material Transactions JIM* 54 (9): 1725–1734.

Kiriyama, T., H. Sasaki, A. Hashimoto, S. Kaneko, and M. Maeda. 2014. "Size Dependence of the Drying Characteristics of Single Lignite Particles in Superheated Steam." *Metallurgical and Materials Transactions E* 1 (4): 349–363. doi:10.1007/s40553-014-0037-2

Kiriyama, T., H. Sasaki, A. Hashimoto, S. Kaneko, and M. Maeda. 2016. "Evaluation of Drying Rates of Lignite Particles in Superheated Steam Using Single-Particle Model." *Metallurgical and Materials Transactions E* 3 (4): 308–316. doi:10.1007/s40553-016-0096-7

Kudra, T., and A. S. Mujumdar. 2009. *Advanced Drying Technologies*. 2nd ed. Boca Raton: CRC Press. doi:10.1201/9781420073898

Kwapinski, W., and E. Tsotsas. 2006. "Characterization of Particulate Materials in Respect to Drying." *Drying Technology* 24 (9): 1083–1092. doi:10.1080/07373930600778155

Law, C. L., and A. S. Mujumdar. 2014. "Fluidized Bed Dryers." In *Handbook of Industrial Drying*, 4th ed., 161–186. CRC Press: Taylor & Francis Group.

Le, K. H., T. T. H. Tran, N. A. Nguyen, and A. Kharaghani. 2020. "Multiscale Modeling of Superheated Steam Drying of Particulate Materials." *Chemical Engineering and Technology* 43 (5): 913–922. doi:10.1002/ceat.201900602

Levy, A., and I. Borde. 2001. "Two-Fluid Model for Pneumatic Drying of Particulate Materials." *Drying Technology* 19 (8): 1773–1788. doi:10.1081/DRT-100107272

Li, J., Q.-C. Liang, and L. Bennamoun. 2016. "Superheated Steam Drying: Design Aspects, Energetic Performances, and Mathematical Modeling." *Renewable and Sustainable Energy Reviews* 60: 1562–1583. doi:10.1016/j.rser.2016.03.033

Li, Y. B., J. Seyed-Yagoobi, R. G. Moreira, and R. Yamsaengsung. 1999. "Superheated Steam Impingement Drying of Tortilla Chips." *Drying Technology* 17 (1–2): 191–213.

Looi, A. Y., K. Golonka, and M. Rhodes. 2002. "Drying Kinetics of Single Porous Particles in Superheated Steam under Pressure." *Chemical Engineering Journal* 87: 329–338.

Makkawi, Y. T., and R. Ocone. 2009. "Mass Transfer Coefficient for Drying of Moist Particulate in a Bubbling Fluidized Bed." *Chemical Engineering and Technology* 32 (1): 64–72. doi:10.1002/ceat.200800483

Mujumdar, A. S. 2006. "An Overview of Innovation in Industrial Drying: Current Status and R&D Needs. In: Drying of Porous Materials." In *Drying of Porous Materials*, edited by S. J. Kowalski, 3–18. Dordrecht: Springer. doi:10.1007/978-1-4020-5480-8_2

Mujumdar, A. S. 2014a. *Handbook of Industrial Drying*. Edited by A. S. Mujumdar. 4th ed. New York: CRC Press: Taylor & Francis Group.

Mujumdar, A. S. 2014b. "Introduction, Classification and Selection of Dryers." In *Handbook of Industrial Drying*, edited by A. S. Mujumdar, 4th ed., 25–55. New York: CRC Press: Taylor & Francis Group.

Mujumdar, A. S. 2014c. "Superheated Steam Drying." In *Handbook of Industrial Drying*, edited by A. S. Mujumdar, 4th ed., 421–432. New York: CRC Press: Taylor & Francis Group.

Ortega-Rivas, E. 2012. "Characterization and Processing Relevance of Food Particulate Materials." *Particle & Particle Systems Characterization* 29 (3): 192–203. John Wiley & Sons. doi:10.1002/PPSC.201100016

Patel, S. K., and M. H. Bade. 2020. "Superheated Steam Drying and Its Applicability for Various Types of the Dryer: The State of Art." In *Drying Technology*. Bellwether Publishing. doi:10.1080/07373937.2020.1847139

Rao, Z., Y. Zhao, C. Huang, C. Duan, and J. He. 2015. "Recent Developments in Drying and Dewatering for Low Rank Coals." *Progress in Energy and Combustion Science* 46: 1–11. doi:10.1016/j.pecs.2014.09.001

Sehrawat, R., P. K. Nema, and B. P. Kaur. 2016. "Effect of Superheated Steam Drying on Properties of Foodstuffs and Kinetic Modeling." *Innovative Food Science & Emerging Technologies* 34: 285–301. doi:10.1016/j.ifset.2016.02.003

Sesso, M. L., and G. V. Franks. 2017. "Fracture Toughness of Wet and Dry Particulate Materials Comprised of Colloidal Sized Particles: Role of Plastic Deformation." *Soft Matter* 13 (27): 4746–4755. doi:10.1039/c7sm00814g

Stokie, D., M. W. Woo, and S. Bhattacharya. 2013. "Comparison of Superheated Steam and Air Fluidized-Bed Drying Characteristics of Victorian Brown Coals." *Energy and Fuels* 27 (11): 6598–6606. doi:10.1021/ef401649j

Tatemoto, Y., S. Yano, T. Takeshita, K. Noda, and N. Komatsu. 2008. "Effect of Fluidizing Particle on Drying Characteristics of Porous Materials in Superheated Steam Fluidized Bed under Reduced Pressure." *Drying Technology* 26 (2): 168–175. doi:10.1080/07373930701831267

Van't Land, C. M. 2012. *Drying in the Process Industry*. *John Wiley*. 1st ed. Hoboken, New Jersey: John Wiley & Sons.

Woo, M. W., D. Stokie, W. L. Choo, and S. Bhattacharya. 2013. "Master Curve Behaviour in Superheated Steam Drying of Small Porous Particles." *Applied Thermal Engineering* 52 (2): 460–467. Elsevier. doi:10.1016/j.applthermaleng.2012.11.038

Zhu, J., Q. Wang, and X. Lu. 2015. "Status and Developments of Drying Low Rank Coal with Superheated Steam in China." *Drying Technology* 33 (9): 1086–1100. doi:10.1080/07373937.2014.942914

10 Miscellaneous Drying Techniques for Particulates

Chung Lim Law
Department of Chemical and Environmental Engineering, University of Nottingham Malaysia, Semenyih, Selangor, Malaysia

Shivanand Shankarrao Shirkole
Department of Food Engineering and Technology, Institute of Chemical Technology Mumbai, ICT – IOC Odisha Campus, Bhubaneswar, India

Sachin Vinayak Jangam
Department of Chemical and Biomolecular Engineering, National University of Singapore, Singapore

CONTENTS

DOI: 10.1201/9781003207108-10

10.1 TRADITIONAL DRYING TECHNIQUES FOR PARTICULATES

The particulate materials are ubiquitous and are widely manufactured for improving the versatility of solids. Seeds and grains are natural particulate materials. Commercially, particulates are prepared by size reduction of coarser solids to improve the product's dispersibility and consistency (Keey, 1991). However, these materials often contain moisture, which must be removed to improve stability and particle flowability. The particulates are smaller in size and need specially designed dryers equipped for dehydration. Several design considerations are incorporated into the traditional dryer for improving the drying efficiency and product quality (A. Mujumdar, 2001; Kwapinski and Tsotsas, 2006). Generally, the desiccation is aided by the forced removal of moisture from particulates through heat or pressure. The heat is often added by an external source via conduction, convection, and radiation. Wherein convection is a commonly used mode of heat transfer for dehydrating particulates.

Drying is energy-intensive and represents 7%–15% of the total energy used in the industry (Kudra, 2004). Several efforts and design modifications were thus introduced onto traditional driers to mitigate the operational cost of driers. These modifications were incorporated to enable the traditional driers to handle moist solids by suspending, fluidizing, or cascading.

10.1.1 Traditional Particulate Drying

Conventionally, particulate drying was performed for agricultural products such as grain and seeds. These are generally conducted in small quantities in batch mode. A widely adopted technique is exposed heap drying, where the particulates heap is exposed to sun and wind for drying. Nevertheless, commercially this technique is uncommon due to its reliance on weather conditions and hygiene constraints (Barbosa de Lima et al., 2014). These concerns were removed by designing an indoor dryer equipped with heating elements and blowers. These indeed improved the reliability of the dryer, whereas the product quality, drying rate, and efficiency of the dryers were compromised (Mujumdar, 2001). Tray dryers were introduced to improve the drying rates, considering the limitations. But labour-intensive batch processes and poor product quality are the downsides of the dryer. Replacing the trays with drums or with conveyors would assist the dryer to work in continuous mode and reduce the labour requirements (Delgado and de Lima, 2014).

Agitating pan dryers are also used for dehydrating particulates, where a rotating shaft attached to scrapers along the bottom of the pan would turn the product over during drying. This could facilitate a uniform and agglomerated product. Besides, integrating these technologies with a vacuum generation system would make the dryer suitable for dehydrating heat-sensitive materials. Nonetheless, these dryers were not continuous dehydrators and are often associated with longer drying time and poor drying efficiency. This led to the development of fluidised dryers, where the particulates were fluidised by hot arid air. Fluidising the particulate during dehydration ascribed faster drying with better product quality. Meanwhile, spray dryers were also introduced for dehydrating particulates from a liquid feed. Spray driers utilize an atomizer or a nozzle to nebulize the liquid feed into a high temperature dry air chamber. This would facilitate the individual liquid droplets being dried into particulates, leading to better quality end product (A. Mujumdar, 1994).

10.1.2 Commercial Particulate Dehydrator

It is important to note that these conventional dryers are not equipped to handle a wider variety of feed. It could be observed that each dryer performs differently and is suitable for specific feeds. Thus, differentiating these dryers according to the feed types would help in a better classification (A. Mujumdar, 2001). The dryers hence could be differentiated according to the suitable feed type as follows.

10.1.2.1 Liquid Feed Dryer

Liquid materials are frequently dried to produce particulates. A drum dryer is a conventional technology used for dehydrating liquid feeds. It is recommended to concentrate the liquid feed to a slurry prior to drying. The product is fed onto a hot drum, and the dehydrated product is scraped off the drum. This results in a flaky, low porous, and high bulk density product. However, drum dryers are currently being replaced by spray dryers. As mentioned earlier, individual drying of droplets will result in a product with a desired particle size and distribution. After dehydration, the particulates are conveyed by the dry air and separated in a cyclone separator or a baghouse. Besides these, a fluidized bed, vibrating bed, and pulse combustion dryers are also used.

10.1.2.2 Sludge or Paste Dryers

Depending on the viscosity and solid content, several dehydration strategies could be adopted for paste or slurry feeds. Thin pastes could be dried using a drum dryer and fluidized bed dryers containing inert particles. The thick paste could be diluted into a pumpable slurry and dehydrated in a liquid feed dryer. Otherwise, the thick paste could be converted to pellets and dehydrated using conventional tray dryers, conveyor dryers, fluidized bed dryers, flash dryers, and spouted bed dryers.

10.1.2.3 Wet Particulate Dryer

Several dryers are available for dehydrating wet particulate dyers. These dryers differ on the dwell time for drying. The selection of the dryers is performed by considering the characteristics of the feed. Rotary dryers are often used for wet particulates, and they are attributed to high throughput and versatility. However, heavy capital investment offset this dryer and is not suitable for cohesive materials. Axial gas flow dryers are also used, and higher efficiencies and turndown ratios attribute them. Moreover, a fluidized/vibrating bed dryer can substitute for drying sticky particulates. Specifically, flash dryers could dehydrate non-porous products, where the surface moisture is only removed.

Thus, the selection of dryers should be performed by considering the form of the feeds and the characteristics of particulate. Each dryer is ascribed to several advantages and limitations. These limitations can be offset by the advantages and could be overcome by minor design considerations and pre-processing methods.

10.2 INNOVATION IN DRYING OF PARTICULATE MATTER

As mentioned in the previous section, there are several dryers that have been built and used successfully over the decades for drying particulates. Although drying technology is a mature research field, there lies several opportunities for innovation on various fronts. The conventional dryers have certain limitations e.g., non-uniform product quality (which can be result of over drying or under-drying caused by several factors owing to the design and operation of a dryer), long drying times resulting in high energy consumption, high carbon footprint, and high operating cost. Dryer safety is another important issue, specifically in drying particulates. There have been some instances of fire and explosions in drying particulates using conventional fluidized beds and spray dryers.

Several attempts have been made over the years to overcome the operational and design problems or difficulties of conventional dryers, as well as to improve the quality of the dried products. These include use of unconventional drying mediums, creative ways of changing operating conditions, modification of existing dryer designs, or development of completely new drying techniques to address the above-mentioned issues.

The new developments in dryers and evolving drying technologies can be classified in the following categories. The key for success and acceptance of new drying techniques is the cost-effectiveness. New or innovative drying techniques are primarily needed for the following plausible reasons:

- Better product quality
- Drying of new products
- Better thermal efficiency and capacity than current technology permits
- Reduced environmental impact (reduced carbon footprint, use of renewable energy sources)
- Improving safety of dryers

- Shorter drying time compared to conventional dryers
- Lower cost (capital, operational, and maintenance cost)

This chapter will focus on selected innovative drying techniques that have addressed some of the aspects mentioned above. Some of these dryers have been used commercially. The chapter will also summarize innovative dryers that are going through the research and development stage.

10.3 CLASSIFICATION OF INNOVATIVE DRYERS

Figure 10.1 shows the simplest way to classify innovative dryers based on the way heat is supplied, mode of operation, and number of stages.

Figure 10.2 shows a more detailed classification of recent developments in drying particulates. The authors have categorized these dryers based on possible modifications to the conventional dryers, novel drying techniques, and hybrid or multi-stage drying techniques. Most of these ideas have been implemented either on a pilot scale or industrial scale drying. Besides this, Table 10.1 is a summary of selected variants of conventional dryers. Note that the novel dryers can be categorized based on several other criteria.

It is worth nothing that significant advancement has been made over the past few years, which have led to new developments in drying. The selection of drying technique for a particular application is an art and depends on several criteria. Readers should refer to the handbook of industrial drying for detailed selection guidelines. Only selected innovative drying techniques are briefly discussed in the following sections.

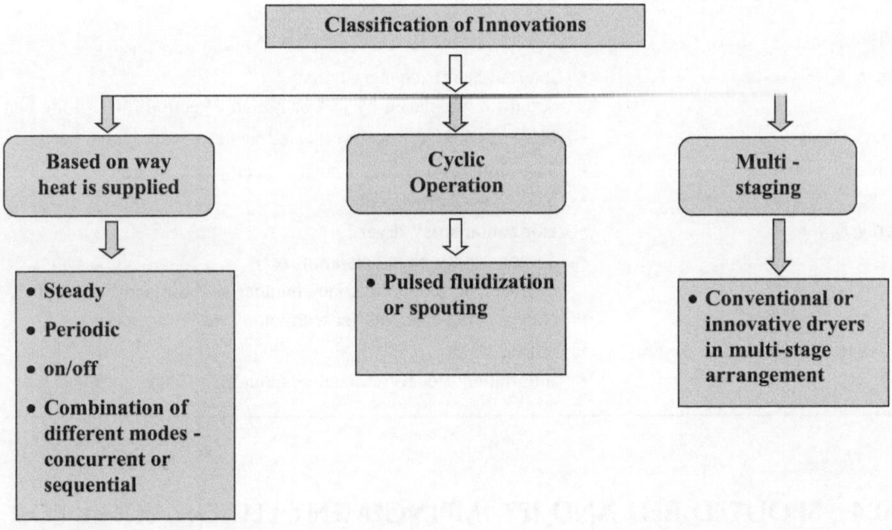

FIGURE 10.1 Classification of innovations in particulate drying.

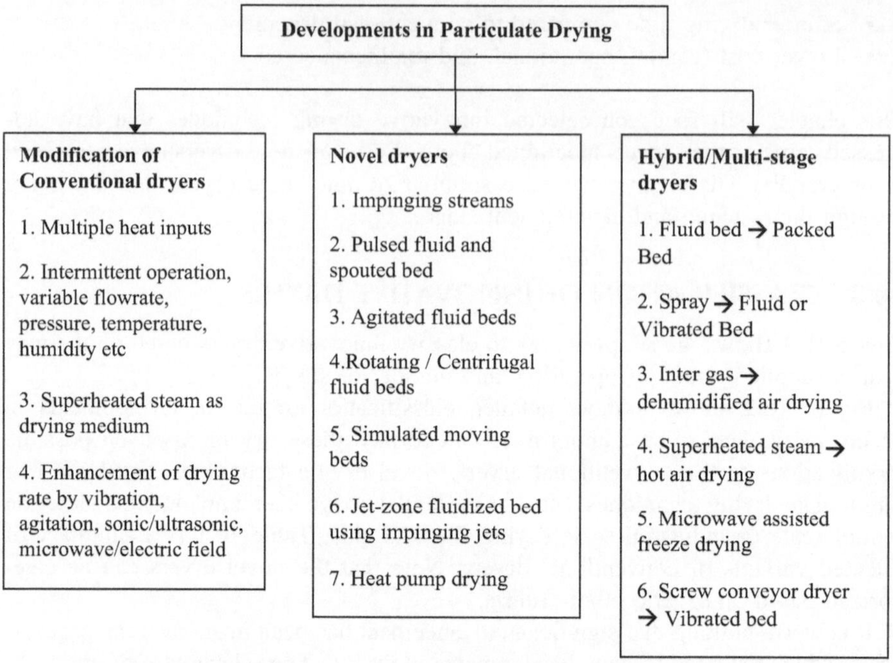

FIGURE 10.2 Development in particulate drying.

TABLE 10.1
Variants of Conventional Dryers

Type	Variants
Rotary	• Internal heat exchanger coils
	• Axial flow replaced by jets of hot air injection into rolling bed
Nauta Dryers	• Planetary mixer; vacuum; heated jacket + microwave heating
Impinging Streams	• Two dimensional; multi-stage; superheated steam
	• Minimize scale-up issue
Spray Dryer	• Horizontal spray dryer
	• Various spray chambers/atomizer
	• Cylinder-on-parabolic cone chamber to minimize wall deposits
	• Nano-spray dryer; ink-jet technology to gentle spray
Fluid Bed/Spouted Bed Dryers	• Pulsed flows
	• Intermittent, local fluidization/spouting

10.4 SPOUTED BED AND JET IMPINGEMENT FLUIDIZATION FOR DRYING PARTICULATES

Spouted bed dryer (SBD) shows various advantages and a few limitations as well compared to conventional dryers. SBDs can used for drying heat-sensitive solids or particulates such as pharmaceuticals, foods, and plastics as SBDs provide short dwell time in the spout.

10.4.1 Spouted Bed Drying

The need for higher circulation and better contact between the drying medium and matter to be dried was the advent cause for the use of spouted bed drying in the food industry. Since then, spouted bed drying has been widely applied for the drying of various food commodities and pharmaceuticals formulation (de Freitas, 2019). It can be applied for the drying matter with characteristic constant rate and falling rate periods with the use of inert material further drying of slurries and paste is also achieved (Pallai, Szentmarjay, and Mujumdar, 2020). The spouted bed can be further segregated into annulus, spout, and fountain for different drying conditions of spouting these regions are either in a packed or fluidized state. The classical model of spouted bed dryer is improved with the draft tubed spouted bed drying with improved capacity and scalability. Adding to that, de Brito et al. (2021) proved that draft tubes spouted bed dryers can reduce the energy issues associated. Reducing the overall energy consumption, and improving the efficiency of the process to achieve a sustainable process is the requirement of current industrial processes, spouted bed drying is found to have all these characteristics owing to the higher heat and mass transfer rate. It is similar to the fluidized bed dryer such that the pressure drop increases up to a certain velocity and remains constant; however, in the case of spouted bed drying, the pressure drop increases suddenly due to bubbling.

The use of inert particles is a common practice in spouted bed dryers; however, it's worth noting that the use of inert particles influences the drying time, rate of drying, as well as energy consumption. de Brito et al. (2021) observed when 25% of inert material was used the drying time was reduced by 25%. Further use of intermittent drying can enable a reduction of energy consumption by up to 67% (Brito, Béttega, and Freire, 2019). The application of intermittent drying mode to spouted bed drying is coming in trend nowadays with various promising results. It has been found that an intermittent drying mode in spouted bed dryers precisely minimizes the resources and results in the product with minimum effect on the quality. Recently, intermittent spouted bed drying is used for soybean seeds, fruit pulp, and dairy-related products (T. N. P. Dantas et al., 2018; S. C. d. M. Dantas et al., 2019; Brito, Zacharias, Forti, and Freire, 2021).

The application of spouted beds in hybrid mode with other technologies also has the potential to reduce the drying time, and generate a product with more retention of volatiles, better quality characteristics, and phenolic compounds. The spouted bed dryer can work complementary with other methods to overcome their limitation as seen in combination with the microwave drying, use of pulsed spouted microwave freeze dryer resulted in a uniform product temperature of the apple cuboids and natural color (D. Wang, Zhang, Wang, and Martynenko, 2018). This is attributed to pulsed spouting applied during the drying operation, increasing the effective diffusion coefficient and heat mass transfer coefficient. A similar trend was reported for the drying of carrageenan (Serowik et al., 2018). The retention of

TABLE 10.2
Drying of Different Food Products Using Spouted Bed Dryer

Food Commodity	Drying Conditions	Effect of Spouted Bed Dryer	Reference
Apple cuboids	640 W, 90 min (0.5 s/min), 640 W, 90 min (0.5 s/min), 640 W, 135 min (0.5 s/10 min) (MW assisted pulsed spouted bed.	Preservation of colour and volatiles	(D. Wang et al., 2018)
Rosemary powder extract	80°C, 150°C, ratio of mass feed flow rate & evaporation rate: 15%, 45%, 75%	Higher retention of total phenol and flavonoids	(Souza et al., 2020)
Carrageenan	100°C, 1.5–4.5 m/s, (Microwave assisted: 0.5 W/g)	Reduction in the bulk density, no effect on the solubility and rheological properties	(Serowik et al., 2018)
Black pepper	45°C–75°C, 1.6, 1.97, 2.37 m/s	Increase in diffusion coefficient	(Jayatunga and Amarasinghe, 2019)
Milk powder	70°C, 2.5 m/s	Higher powder production rate, thermal efficiency and yield.	(Vieira, Olazar, Freire, and Freire, 2019)
Flaxseeds	40°C, 60°C, 80°C,	Lower peroxide value	(Dehghan-Manshadi, Peighambardoust, Azadmard-Damirchi, and Niakousari, 2020)

heat labile is an unarguable property of the spouted bed dryer, Souza et al. (2020) found a maximum 12.3 to 4.7 times higher retention of total phenols and flavonoids in the spouted bed dried extract for the rosemary powder than the conventional extract. Mechanical damage to the feed due to agitation is one of the shortcomings of this method (Brito et al., 2021). Table 10.2 gives the different food commodities for which spouted bed drying is applied in recent years and their corresponding effects on the dried food product or drying operation.

10.4.2 Air Impingement Drying

It is quite a well-known fact that the drying of the food material is achieved by internal diffusion and external mass transfer. The use of impingement in the drying methods comes into play as an aid to facilitate the latter step in drying. The implementation can be commonly achieved by air or steam, superheated steam can also be applied, and it can be applied in pulsed or continuous mode. Apart from the improved external mass transfer, the impingement drying has shown various other beneficial effects in terms of the product quality achieved. Impingement is usually applied in conjugation or in series with other drying methods like rotary, hot air hybrid mode with other dryers, microwave, vacuum, and so on where it shows improved heat transfer rate. The method works best when the food commodity is high in unbound water, a substantial deteriorative effect on the quality of the product is evident in the case of low unbound food (A. S. Mujumdar and Huang, 2020). The jet temperature and jet velocity are the critical parameters of the drying conditions. J. Wang et al. (2020), based on an investigation, found the jet temperature to have a higher influence on the heat transfer rate than the jet velocity on the heat transfer rate.

Food high in surface or unbound moisture can be easily dried by this method followed by the other conventional method of drying. This will not only reduce the overall energy consumption and improve efficacy but also deliver a product with superior quality. The reduction in the drying time due to improved heat transfer rate is the vital feature of impingement drying that has raised the use of these techniques. A marked reduction in the drying of air-impinged broccoli florets was noted; on the same lines, a 33% reduction in the drying time was recorded for the drying of apple slices (Liu et al., 2019; Peng et al., 2019). A recent study investigated the effect of impingement drying while germination of rice, and the results concluded the increased rate of drying, with higher energy efficiency and less loss of chemical constituents (Netkham, Tirawanichakul, Khummueng, and Tirawanichakul, 2022).

However, the application of this method can be detrimental to the products with lower unbound food, as suggested by A. S. Mujumdar and Huang (2020). Nevertheless, recent studies of impingement drying show a product with superior quality. Apple slices died with air impingement in first stage followed by hot air radio frequency drying retained higher phenolic content, vitamin C and better colour characteristics with 33% less drying time than hot air radio frequency drying (Peng et al., 2019). Further, one of the studies shows that application of air impingement to the drying of the apple slices resulted in higher retention of vitamin C than by hot air and microwave drying methods. A similar trend was seen for phenolic content with even higher retention of phenolic content than the freeze-dried samples (Yin et al., 2019). With reference to the preservation of the flavour and aroma of dried products, impingement is found to be promising than the mere use of conventional techniques for drying. The effect of an air impingement dryer on different food commodities is summarised in Table 10.3.

TABLE 10.3
Drying of Different Food Products Using Air Impingement Dryer

Food commodity	Conditions	Effect of impingement	Reference
Chillies	70°C, 80°C, 90°C, 5, 10, 15 m/s	Accelerated drying rates	(J. Wang et al., 2020)
Apple slices	60°C, 75°C, 1.3–2.3 m/s	Higher retention of nutrients, lower drying time	(Peng et al., 2019)
Apple slices	55°C, 60°C, 65°C, 1.3, 1.8, 2.3 m/s	Better retention of phenolic content and vitamin C	(Yin et al., 2019)
Purple potatoes slices	50°C, 65°C, 80°C, 2.3 m/s	Lower temperature air impingement resulted in higher retention of antioxidant activity	(Qiu et al., 2018)
Broccoli florets	65°C, 9 m/s	Reduction in drying time, higher vitamin C, rehydration ratio	(Liu et al., 2019)
Poria cocos wolf cubes	65°C– 85°C	High-quality product with efficient drying	(Zhang, Chen, Pan, and Zheng, 2021)
Shiitake mushroom	56°C, 9 m/s	Better flavour and inhibition of Maillard reaction	(Luo et al., 2021)

10.5 SCREW CONVEYOR DRYERS

The conventional particulate drying techniques may have one or more constraints, such as large footprint, low heat and mass transfer rates, poor efficiency, non-continuous operation, high cost, suitability for heat-sensitive materials, etc. Screw conveyor dryer (SCD) can overcome much of these limitations. The idea behind a screw conveyor dryer is very simple. It consists of a jacketed conveyor in which material is simultaneously heated and dried as it is conveyed (Figure 10.3). The heating medium can be either hot water, steam, or other suitable high-temperature heating fluid. The shaft of the screw and flight can be made hollow to provide a larger heat transfer area (Waje et al., 2006; Jangam et al., 2011). The mode of heat transfer is mainly conduction; however, a small quantity of drying medium is used to remove the evaporated moisture. SCDs can also have a provision for a vacuum.

The SCD offers relatively high heat transfer area-to-volume ratio compared to other dryers by virtue of the screw geometry that acts as an immersed heat transfer surface (Waje et al., 2006). The improved heat transfer in screw conveyer dryers is a result of rotation of the screw that continuously renews the heating surface. Besides this, agitation of the particle bed improves the temperature and moisture uniformity of the product. Screw conveyor dryers are also safe to operate for most of the products, as heat is supplied via a heating jacket. A risk of fire is substantially reduced since exposure to air is minimal. Some of the key advantages of screw conveyor dryers are indirect heating, possible size reduction during the drying process, safe operation, and high thermal efficiency.

SCD can be used for solid particles ranging from free-flowing to relatively free-flowing and from fine powder to lumpy, sticky, and fibrous materials and to materials that become friable at some stage during the drying process (Waje et al., 2006).

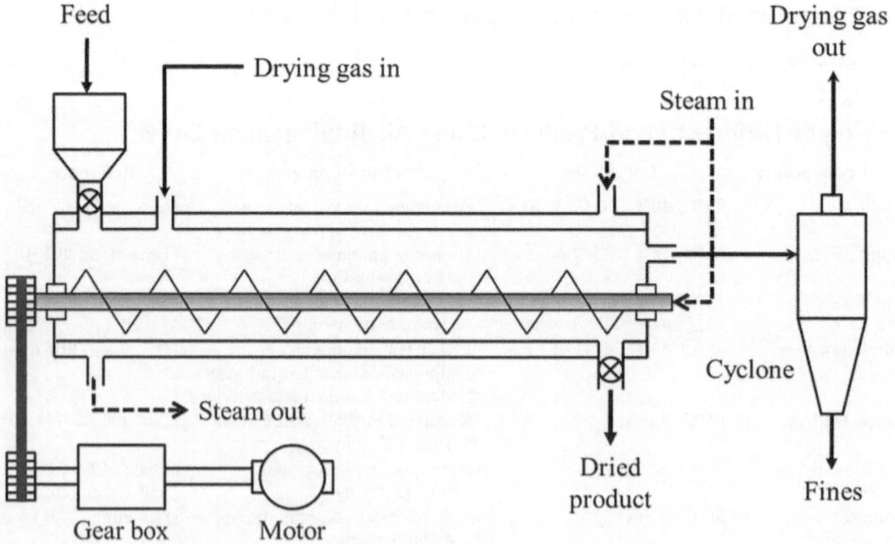

FIGURE 10.3 Schematic of screw conveyor dryer.

There have been several examples of successful pilot scale and industrial scale applications of screw conveyor dryers. These include dehydration of sludge (Kim et al.), drying of low value material such as hog manure (Benali and Kudra,), drying of low rank coal (Jangam et al., 2011), and drying of woodchips (Kaplan and Celik, 2012, 2018).

10.6 MECHANICALLY AGITATED DRYING TECHNIQUES

Conventional drying techniques, such as oven dryer, packed bed dryer, fixed bed dryer, in-bin dryer, and tray dryer, are very common in the processing industry, especially those that handle and dry huge amounts of drying materials. As drying involves heat and mass transfer, therefore the interface between the drying medium (typically drying air) and the drying material is important in order to ensure effective heat transfer and mass transfer. However, stationery dryers mentioned above tend to have the issue of poor contacting efficiency between the drying medium and the drying material; hence, this results in a lower rate of heat transfer and mass transfer. This in turn lowers the drying rate and the drying time is prolonged depending the amount of drying material as well as the characteristics of the drying materials such as size, initial moisture content, etc.

In order to improve the contacting efficiency between the drying medium and the drying material, agitation can be applied to disintegrate the drying materials so that the surface of the drying materials is effectively exposed to the drying medium; hence, facilitating the heat and mass transfer that in turn promotes the drying rate and shortens the drying time. This type of dryer is known as an agitated dryer.

There are many types of agitated dryers. The following are the common mechanically agitated dryers that we can find in the dryer market:

- Agitated pan dryer
- Agitated vacuum dryer
- Agitated conical dryer
- Vertical conical screw dryer
- Horizontal paddle dryer

10.7 AGITATED PAN DRYERS/AGITATED VACUUM PAN DRYER

A pan dryer is a simple dryer where particulate solids are poured into the pan dryer. Most pan dryers are operated in a vacuum in order to lower the drying temperature. In this case, it is known as an agitated vacuum pan dryer. Further, the pan is heated up so that heat is transferred in conduction mode. If the drying materials form a bed of particulate solids in the pan dryer, the issues of poor contact efficiency between the drying material and the drying medium that is mentioned above arises and hence the agitator can be installed to facilitate the disintegration of the particulate solids and the particle flow in the pan dryer. With the installation of a vacuum pump, heater, rotor for the agitator, and temperature controller, the design of an agitated vacuum pan dryer becomes more complex.

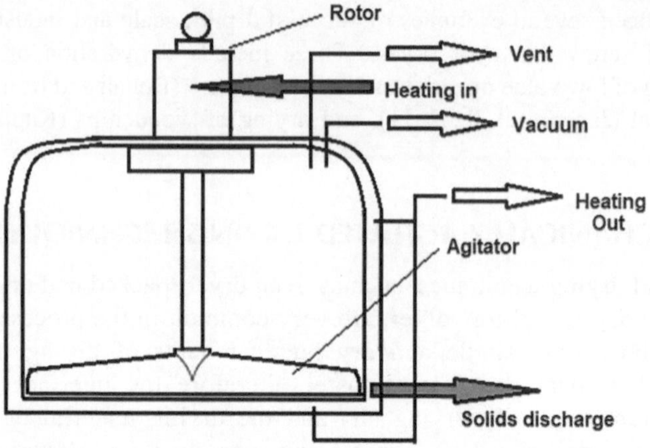

FIGURE 10.4 Agitated vacuum pan dryer.

An agitated vacuum pan dryer is suitable for drying high-risk material such as materials that are flammable or contain toxic solvent or flammable solvent. In this regard, displacement of oxygen can be achieved by carrying out inertizing using nitrogen that is charged into the pan dryer chamber. Then, the drying material is poured into the pan dryer chamber while the dryer wall and agitator are heated up. Depending on the drying temperature, the heating medium can be hot water or steam. In the case of containing toxic solvent, a condenser can be installed to capture the toxic solvent from the drying process and condense it into liquid and contain it in a container. Figure 10.4 shows the schematic diagram of an agitated vacuum pan dryer.

10.7.1 Agitated Vacuum Dryers

The working principle of an agitated vacuum dryer is like an agitated vacuum pan dryer. Instead of using a pan dryer chamber, other types of dryer chamber design may be applied as long as the dryer chamber has an agitator, and the drying chamber can be operated at vacuum condition.

10.7.2 Agitated Conical Dryers/Agitated Conical Screw Dryers

Likewise, agitated conical dryers and agitated conical screw dryers operate using the same operating principle as agitated vacuum dryers. For agitated conical dryers, the unique feature is the shape of the dryer chamber, which is conical in shape and its agitator is designed in a way to suit the conical shape of the conical dryer. The agitated conical screw dryer has a screw-type agitator that is rotated about its own axis and revolves along the walls of the conical dryer chamber.

10.7.3 Agitated Filter Dryer

An agitated pan dryer can be combined with a filter and becomes a multipurpose unit operation that can perform filtration and washing, followed by drying. This unit

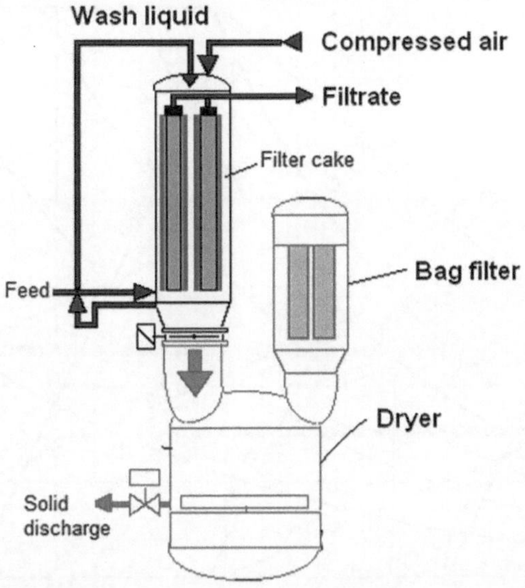

FIGURE 10.5 Schematic diagram of a vertical filter dryer.

is useful to convert solution and slurry into solids and powder. A filter dryer combines mechanical dewatering of a slurry using a filtration technique, followed by thermal drying of the filter cake. After the drying operation, the filter cake is discharged, and the solids can be in the form of powder or particulate solids. The filter dryer is operated in a closed system; hence, it eliminates the issues of solids handling, cleaning, and containment associated with filtration and drying that are carried out in separate unit operations.

Figure 10.5 shows the schematic diagram of a vertical filter dryer. Filtration is carried out in the unit located in the top left. During the filtration operation, the filtrate is withdrawn, and a filter cake is formed with the assistance of compressed air that exerts a driving force for the filtrate to permeate through the filter elements. Washing is performed after the filtration is completed. This is done with the aim to remove impurities in the filter cake. The filter cake is then subjected to thermal drying to remove the surface moisture in the cake. Thereafter, the filter cake is loosened from the filter elements, and it drops into a dryer located beneath the filter dryer. A second drying is carried out to remove the internal moisture. In this regard, an agitated dryer may be considered; else, a fluidised bed dryer is also a good option.

10.8 HEAT PUMP DRYING

Heat pump drying is also known as heat pump assisted drying. It is similar to a hot air dryer; any hot air dryer can be converted into heat pump assisted drying. The main difference between hot air drying and heat pump assisted drying is the drying medium (drying air) that is supplied to the drying chamber. A hot air dryer requires a hot air generator such as combustor or heating element to generate a

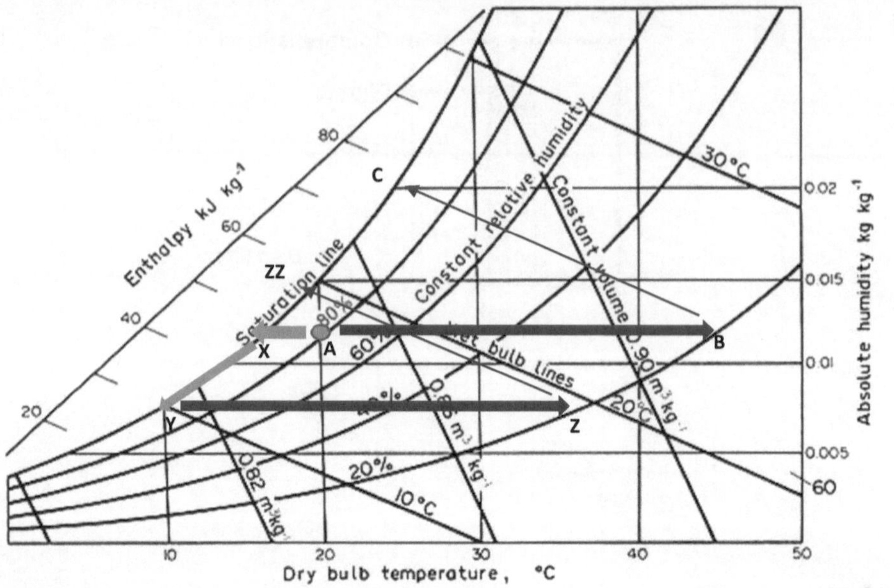

FIGURE 10.6 Psychrometic chart and air pathway undergoing hot air drying operation and heat pump assisted drying operation.

drying medium at an elevated temperature. A heat pump assisted dryer requires a heat pump module to assist in the generation of low-temperature dehumidified air. Figure 10.6 shows a psychrometric chart. Say, atmospheric air near the dryer surrounding is 20°C and its relative humidity is 20%. Point A in Figure 10.6 indicates the location of the atmospheric air on the psychrometric chart. Under hot air drying operation, the atmospheric air would be heated up to 45°C to reach a relative humidity of 20%. This is indicated by the line A – B. Here we can see that the absolute humidity of the air remains the same while the air is heated up. On the other hand, in a typical heat pump assisted drying, the air would be cooled to its saturation point (x), subsequent cooling would result in condensation of water vapour in the air and its absolute humidity is lowered (Y). The dehumidified air is then heated up to 35°C to reach a relative humidity of 20%. The dehumidified air at 35% RH would have a similar drying capacity as hot air at 45% RH. Hence, heat pump assisted drying is classified as a low-temperature drying method.

Under hot air drying operation, the drying air would follow the pathway of B – C when it makes contact with drying materials in the drying chamber. During this process, the drying air picks up moisture from the drying materials and its absolute humidity as well as its relative humidity increase. The drying air moves closer to its saturation point (C).

Under heat pump assisted operation, the dehumidified air would follow the pathway of Z – ZZ. Likewise, the dehumidified air absorbs moisture during drying and hence its absolute humidity and relative humidity increase. The saturation point is ZZ.

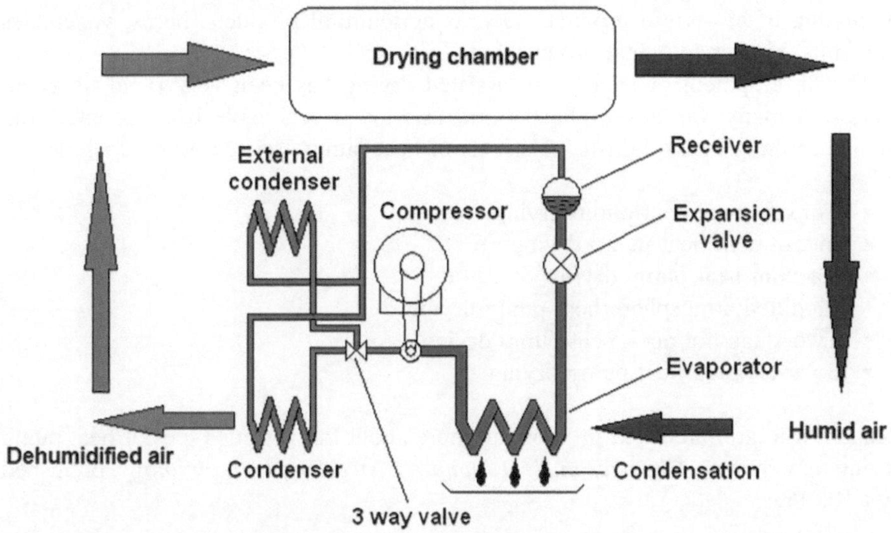

FIGURE 10.7 Heat pump dryer working principles.

Figure 10.7 shows the working principle of heat pump assisted drying. Humid air that is from a drying chamber has a higher humidity due to the water vapour that it carried away from the drying materials in the drying chamber. When it makes contact with the evaporator of a heat pump module, condensation occurs, and water is extracted from the humid air. The dehumidified air then makes contact with the condenser of the heap pump module and its temperature is elevated. Now the drying air has a higher temperature, and its relative humidity is lower due to the heating. The low-temperature dehumidified air is then charged into the drying air chamber for drying and removal of moisture from the drying materials.

Heat pump assisted drying is a good option to dry materials that are sensitive to high temperature (above 60°C) or materials that contain ingredients that are heat sensitive. Therefore, heat pump assisted drying can be considered for the process industry that intend to retain the colour pigments, aromatic compounds, and bio-active ingredients in the processed materials. Hence, it is useful to vegetables, fruits, herbs, medicinal plants, and biotechnological product processing industries.

Heat pump drying is energy efficient as this drying technique allows heat recover from the exhaust air (Rossi et al., 1992; Chua et al., 2002). In addition, it allows easy control of the drying air temperature and humidity (Prasertsan et al., 1996). Researchers have reported that heat pump drying could produce better quality of bio-origin products, which include the type of materials mentioned above (Prasertsan et al., 1998; Adapa et al., 2002; Alves-Filho, 2002; Sosle et al., 2003). In addition, researchers have also reported that heat pump assisted drying operated in a closed system could avoid the release of gases and fumes into the atmosphere as well as preventing product contamination (Perera and Rahman, 1997; Colak and Hepbasli, 2009; Chua et al., 2010; Hii et al., 2012; Patel and Kar, 2012). Hence, this confirms that heat pump assisted drying is a good option for the drying and processing of bio-origin products such as agricultural products, herbs, vegetables and fruits, biotechnological products, etc.

The development of heat pump assisted drying has been very rapid in recent years and many variants of heat pump assisted dryers have been adopted and applied in the process industry. Variants of heat pump assisted dryers include:

- Conventional heat pump drying
- Intermittent heat pump drying
- Vacuum heat pump drying
- Modified atmosphere heat pump drying
- Two-stage hot air – heat pump drying
- Solar-assisted heat pump drying

Readers who are interested in knowing more about the various types of heat pump drying may refer to *Advances in Heat Pump-Assisted Drying Technology*, published by CRC Press.

10.9 MICROWAVE-ASSISTED DRYING

Microwaves are wavelengths/radio frequencies ranging from 300 MHz–300 GHz. They are mostly shorter than infrared and longer than radio waves. Microwave-assisted drying involves the utilization of dielectric heating as a source of thermal energy for the removal of moisture thereby reducing the drying time. It is basically a two-step process where polarization and depolarization of ions are induced when the alternating electromagnetic wave interacts with the material. This generates volumetric heating that converts the moisture inside the product into vapors. The step following this requires immediate removal of vapor that takes place due to the vapor pressure gradient on application of air over the surface. This technology is considered the first and foremost method involving a convective mode of heating rather than conductive mode. The mechanism of energy dissipation of microwaves highly depends on ionic conductivity, Maxwell–Wagner polarization, free water polarization, bound water polarization, etc. (Tang, Feng, and Lau, 2002). Microwave drying has been applied in various fields like drying metallic powders, bulk metals, and composite materials to edibles. Based on its application, microwave processing can be in three different segments (Mishra and Sharma, 2016):

a. High-temperature processing involves usage of microwaves above 1,000°C. Applied mostly for ceramics with higher density, joining of bulk metals, etc.
b. Medium-temperature processing materials are processed in the range of 500°C–1,000°C. Specially used for melting of glass, heating of metallic powders, etc.
c. Low-temperature processing is carried out at temperatures >500°C. This method finds its application especially in textile, rubber, and food industries for synthesis and drying.

10.9.1 Parameters Influencing Microwave Drying

There are many parameters that influence the microwave-assisted drying; however, a few parameters are discussed shortly as follows.

10.9.1.1 Dielectric Property

Microwave-assisted drying works best with materials exhibiting excellent dielectric properties (Rattanadecho and Makul, 2016). Dielectric properties of materials are generally governed by various parameters such as temperature applied, frequency, particle size density, and most importantly the availability of moisture in the material.

10.9.1.2 Moisture

Moisture content of the product subjected to microwave-assisted drying is an important parameter influencing the efficiency of the process. The dielectric constant of a material (Ɛ") determines the heat generated on application of the electromagnetic waves. Since Ɛ" is also a function of its polarity, therefore, products with higher levels of moisture will absorb more microwave energy, thereby leading to a rapid rate of drying.

10.9.1.3 Microwave Energy Intensity

Microwave intensity is positively correlated to drying rate; however, in the case of polymers that are susceptible to thermal runaway phenomenon (sudden increase in Ɛ" with increase in temperature), higher power intensity may increase the chances of product damage, especially during the end of the drying stage where the evaporative cooling effect is very minimal. An increase in amount of product exposed to drying at a high intensity of microwave energy might resolve the use of product damage due to thermal runaway.

10.9.2 Advantages and Drawbacks of Microwave-Assisted Drying

The advantages and drawbacks associated with microwave-assisted drying are discussed shortly as follows.

10.9.2.1 Advantages

Microwave-assisted drying provides an edge over the other conventional drying technologies by enhancing the drying rate, thereby improving the energy efficiency. Microwave energy can penetrate deep inside the matrices on which it is used, thereby improving the heating efficiency. It also utilizes non-polluting energy sources, involves lower processing cost, and is also a self-regulating system.

10.9.2.2 Drawbacks

The non-uniform heating behaviour of a microwave is one of the major disadvantages of microwave drying. Thermal damages incurred due to this setback can be avoided if the product exposed to microwave drying is in constant motion. Another disadvantage is penetration depth. Even though microwaves can penetrate greater depths of the matrices provided for drying, they are still minimal compared to that achieved using radio frequencies.

10.9.3 Combined Application of Microwave Drying

Microwaves have been reported to be used in various combinatorial applications for the purpose of drying and some of the most commonly reported methods have been discussed shortly as follows.

10.9.3.1 Microwave-Assisted Vacuum Drying

The vacuum drying process involves drying with reduced pressure, which lowers the boiling point of water and thereby enhances the drying rate. However, transmission of energy in the absence of air is a limitation for this process. Use of microwaves for energy dissipation into the system to be dried helps in overcoming this disadvantage. Moreover, due to the absence of air drying, this method is also advantageous to maintain the colour characteristics and oxidative browning of the dehydrated product.

10.9.3.2 Microwave-Assisted Hot Air Drying

Conventional hot air drying is the most widely adopted drying method on a commercial scale. However, this method incurs several disadvantages, such as longer drying time, case hardening of the dried product, minimum volume retention, etc. Application of microwaves along with air drying, the drying rate is increased, thereby reducing the drying time.

10.9.3.3 Microwave-Assisted Freeze Drying

A major disadvantage of the freeze-drying process is that due to the availability of moisture in the frozen state, the transmission of water from deep matrices to the surface can take place only by diffusion, which limits the drying rate. So, involving the microwave for volumetric heating helps in overcoming the drawback. But a major issue related to microwave-assisted freeze drying is that due to non-uniform heating of the microwave, the product might undergo thermal runaway, affecting the quality of the product.

REFERENCES

Adapa, P. K., Sokhansanj, S., & Schoenau, G. J. (2002). Performance study of a re-circulating cabinet dryer using a household dehumidifier. *Drying Technology* 20(8): 1673–1689.

Alves-Filho, O. (2002). Combined innovative heat pump drying technologies and new cold extrusion techniques for production of instant foods. *Drying Technology* 20(8): 1541–1557.

Brito, R., Béttega, R., & Freire, J. (2019). Energy analysis of intermittent drying in the spouted bed. *Drying Technology* 37(12): 1498–1510. doi:10.1080/07373937.2018.1512503

Brito, R., Zacharias, M., Forti, V., & Freire, J. (2021). Physical and physiological quality of intermittent soybean seeds drying in the spouted bed. *Drying Technology* 39(6): 820–833. doi:10.1080/07373937.2020.1725544

Chua, K. J., Chou, S. K., Ho, J. C., & Hawlader, M. N. A. (2002). Heat pump drying: Recent developments and future trends. *Drying Technology* 20 (8): 1579–1610.

Chua, K. J., Chou, S. K., & Yang, W. M. (2010). Advances in heat pump systems: A review. *Applied Energy* 87(12): 3611–3624.

Colak, N., & Hepbasli, A. (2009). A review of heat-pump drying (HPD): Part 2 – Applications and performance assessments. *Energy Conversion and Management* 50(9): 2187–2199.

Dantas, T. N. P., Moraes Filho, F. C., Souza, J. S., Oliveira, J. A. D., Rocha, S. C. D. S., & Medeiros, M. D. F. D. D. (2018). Study of model application for drying of pulp fruit in spouted bed with intermittent feeding and accumulation. *Drying Technology* 36(11): 1349–1366. doi:10.1080/07373937.2017.1402785

Dantas, S. C. D. M., Pontes Júnior, S. M. D., Medeiros, F. G. M. D., Santos Junior, L. C., Alsina, O. L. S. D., & Medeiros, M. D. F. D. D. (2019). Spouted-bed drying of acerola pulp (Malpighia emarginata DC): Effects of adding milk and milk protein on process performance and characterization of dried fruit powders. *Journal of Food Process Engineering* 42(6): e13205. doi:10.1111/jfpe.13205

de Brito, R. C., Tellabide, M., Estiati, I., Freire, J. T., & Olazar, M. (2021). Drying of particulate materials in draft tube conical spouted beds: Energy analysis. *Powder Technology* 388: 110–121. doi:10.1016/j.powtec.2021.04.074

de Freitas, L. A. P. (2019). Pharmaceutical applications of spouted beds: A review on solid dosage forms. *Particuology* 42: 126–136. doi:10.1016/j.partic.2018.05.002

Dehghan-Manshadi, A., Peighambardoust, S. H., Azadmard-Damirchi, S., & Niakousari, M. (2020). Effect of infrared-assisted spouted bed drying of flaxseed on the quality characteristics of its oil extracted by different methods. *Journal of the Science of Food Agriculture* 100(1): 74–80. doi:10.1002/jsfa.9995

Delgado, J., & de Lima, A. B. (2014). *Transport phenomena and drying of solids and particulate materials* (Vol. 48). Springer.

Hii, C. L., Law, C. L., & Suzannah, S. (2012). Drying kinetics of the individual layer of cocoa beans during heat pump drying. *Journal of Food Engineering* 108(2): 276–282.

Jayatunga, G., & Amarasinghe, B. (2019). Drying kinetics, quality and moisture diffusivity of spouted bed dried Sri Lankan black pepper. *Journal of Food Engineering*, 263, 38–45. doi:10.1016/j.jfoodeng.2019.05.023

Keey, R. (1991). *Drying of loose and particulate materials*. CRC Press.

Kudra, T. (2004). Energy aspects in drying. *Drying Technology* 22(5): 917–932. doi:10.1081/DRT-120038572

Kwapinski, W., & Tsotsas, E. (2006). Characterization of particulate materials in respect to drying. *Drying Technology* 24(9): 1083–1092. doi:10.1080/07373930600778155

Liu, Z.-L., Bai, J.-W., Yang, W.-X., Wang, J., Deng, L.-Z., Yu, X.-L., … Xiao, H.-W. (2019). Effect of high-humidity hot air impingement blanching (HHAIB) and drying parameters on drying characteristics and quality of broccoli florets. *Drying Technology*. doi:10.1080/07373937.2018.1494185

Luo, D., Wu, J., Ma, Z., Tang, P., Liao, X., & Lao, F. (2021). Production of high sensory quality Shiitake mushroom (Lentinus edodes) by pulsed air-impingement jet drying (AID) technique. *Food Chemistry* 341: 128290. doi:10.1016/j.foodchem.2020.128290

Mishra, R. R., & Sharma, A. K. (2016). Microwave–material interaction phenomena: Heating mechanisms, challenges and opportunities in material processing. *Composites Part A: Applied Science and Manufacturing* 81: 78–97. doi:10.1016/j.compositesa.2015.10.035

Mujumdar, A. (1994). *Handbook of industrial drying* (Vol. 1). Taylor & Francis.

Mujumdar, A. (2001). *Recent developments in the drying technologies for the production of particulate materials*. doi: 10.1016/S0167-3785(01)80056-7

Mujumdar, A. S., & Huang, B. (2020). *Impingement drying. Handbook of industrial drying* (pp. 489–501). CRC Press.

Netkham, H., Tirawanichakul, S., Khummueng, W., & Tirawanichakul, Y. (2022). Impingement drying of germinated brown rice varieties at intermediate temperatures: Drying kinetics and analysis of quality. *Journal of Food Nutrition Research*, 61(1).

Pallai, E., Szentmarjay, T., & Mujumdar, A. S. (2020). *Spouted bed drying. Handbook of industrial drying* (pp. 453–488). CRC Press.

Patel, K. K., & Kar, A. (2012). Heat pump assisted drying of agricultural produce – An overview. *Journal of Food Science and Technology* 49(2): 142–160.

Peng, J., Yin, X., Jiao, S., Wei, K., Tu, K., & Pan, L. (2019). Air jet impingement and hot air-assisted radio frequency hybrid drying of apple slices. *LWT* 116: 108517. doi:10.1016/j.lwt.2019.108517

Perera, C. O., & Rahman, M. S. (1997). Heat pump dehumidifier drying of food. *Trends in Food Science and Technology* 8(3): 75–79.

Prasertsan, S., & Saen-saby, P. (1998). Heat pump drying of agricultural materials. *Drying Technology* 16(1 & 2): 235–250.

Prasertsan, S., Saen-saby, P., Nyamsritrakul, P., & Prateepchaikul, G. (1996). Heat pump dryer. Part 1: Simulation of the models. *International Journal of Energy Research* 20: 1067–1079.

Qiu, G., Wang, D., Song, X., Deng, Y., & Zhao, Y. (2018). Degradation kinetics and antioxidant capacity of anthocyanins in air-impingement jet dried purple potato slices. *Food Research International* 105: 121–128. doi:10.1016/j.foodres.2017.10.050

Rattanadecho, P., & Makul, N. (2016). Microwave-assisted drying: A review of the state-of-the-art. *Drying Technology* 34(1): 1–38. doi:10.1080/07373937.2014.957764

Rossi, S. J., Neues, C., & Kicokbusch, T. G. (1992). Thermodynamics and energetic evaluation of a heat pump applied to drying of vegetables. In: Mujumdar, A. S. (Ed.), *Drying'92*. Elsevier Science Publishers, Amsterdam, pp. 1475–1478.

Serowik, M., Figiel, A., Nejman, M., Pudlo, A., Chorazyk, D., Kopec, W., … Rychlicka, J. (2018). Drying characteristics and properties of microwave- assisted spouted bed dried semi-refined carrageenan. *Journal of Food Engineering* 221: 20–28. doi:10.1016/j.jfoodeng.2017.09.023

Sosle, V., Raghavan, G. S. V., & Kittler, R. (2003). Low-temperature drying using a versatile heat pump dehumidifier. *Drying Technology* 21(3): 539–554.

Souza, C. R., Baldim, I., Bankole, V. O., da Ana, R., Durazzo, A., Lucarini, M., … Oliveira, W. P. (2020). Spouted bed dried Rosmarinus officinalis extract: A novel approach for physicochemical properties and antioxidant activity. *Agriculture* 10(8): 349. doi: 10.3390/agriculture10080349

Tang, J., Feng, H., & Lau, M. (2002). Microwave heating in food processing. *Advances in Bioprocessing Engineering* 1: 1–43.

Vasile, Minea (Ed.). (2016). *Advances in heat pump-assisted drying technology*. CRC Press.

Vieira, G. N., Olazar, M., Freire, J. T., & Freire, F. B. (2019). Real-time monitoring of milk powder moisture content during drying in a spouted bed dryer using a hybrid neural soft sensor. *Drying Technology* 37(9): 1184–1190. doi:10.1080/07373937.2018.1492614

Wang, D., Zhang, M., Wang, Y., & Martynenko, A. (2018). Effect of pulsed-spouted bed microwave freeze drying on quality of apple cuboids. *Food Bioprocess Technology* 11(5): 941–952. doi:10.1007/s11947-018-2061-1

Wang, J., Xiao, H.-W., Fang, X.-M., Mujumdar, A., Vidyarthi, S. K., & Xie, L. (2020). Effect of high-humidity hot air impingement blanching and pulsed vacuum drying on phytochemicals content, antioxidant capacity, rehydration kinetics and ultrastructure of Thompson seedless grape. *Drying Technology*, 1–14. doi:10.1080/07373937.2020.1845721

Yin, X., Jiao, S., Sun, Z., Qiu, G., Tu, K., Peng, J., & Pan, L. (2019). Two-step drying based on air jet impingement and microwave vacuum for apple slices. *Journal of Food Process Engineering* 42(5): e13142. doi:10.1111/jfpe.13142

Zhang, W., Chen, C., Pan, Z., & Zheng, Z. (2021). Vacuum and infrared-assisted hot air impingement drying for improving the processing performance and quality of Poria cocos (Schw.) Wolf cubes. *Foods* 10(5): 992. doi:10.3390/foods10050992

Index

Page numbers in **bold** indicate tables.